SpringerBriefs in Applied Sciences and Technology

SpringerBriefs present concise summaries of cutting-edge research and practical applications across a wide spectrum of fields. Featuring compact volumes of 50 to 125 pages, the series covers a range of content from professional to academic.

Typical publications can be:

- A timely report of state-of-the art methods
- An introduction to or a manual for the application of mathematical or computer techniques
- A bridge between new research results, as published in journal articles
- A snapshot of a hot or emerging topic
- An in-depth case study
- A presentation of core concepts that students must understand in order to make independent contributions

SpringerBriefs are characterized by fast, global electronic dissemination, standard publishing contracts, standardized manuscript preparation and formatting guidelines, and expedited production schedules.

On the one hand, **SpringerBriefs in Applied Sciences and Technology** are devoted to the publication of fundamentals and applications within the different classical engineering disciplines as well as in interdisciplinary fields that recently emerged between these areas. On the other hand, as the boundary separating fundamental research and applied technology is more and more dissolving, this series is particularly open to trans-disciplinary topics between fundamental science and engineering.

Indexed by EI-Compendex, SCOPUS and Springerlink.

Beatriz Ledesma Cano · María Alonso Sánchez ·
José Manuel Díaz Rasero · Silvia Román Suero ·
Sergio Nogales Delgado

Introduction to Hydrocarbonization

Principles and Applications

 Springer

Beatriz Ledesma Cano
Department of Applied Physics
University of Extremadura
Badajoz, Spain

María Alonso Sánchez
Department of Applied Physics
University of Extremadura
Badajoz, Spain

José Manuel Díaz Rasero
Department of Applied Physics
University of Extremadura
Badajoz, Spain

Silvia Román Suero
Department of Applied Physics
University of Extremadura
Badajoz, Spain

Sergio Nogales Delgado 🔟
Department of Applied Physics
University of Extremadura
Badajoz, Spain

ISSN 2191-530X   ISSN 2191-5318   (electronic)
SpringerBriefs in Applied Sciences and Technology
ISBN 978-3-031-70038-5   ISBN 978-3-031-70039-2   (eBook)
https://doi.org/10.1007/978-3-031-70039-2

This Springer imprint is published by the registered company Springer Nature Switzerland AG
The registered company address is: Gewerbestrasse 11, 6330 Cham, Switzerland

If disposing of this product, please recycle the paper.

Preface

In recent years, hydrothermal carbonization technology has not only been presented as a promising alternative for the treatment of biomass and other wastes but also as a technique with a strong sustainable approach and aimed at reducing the environmental footprint, with potential applications in areas as diverse as waste management, energy production and soil improvement.

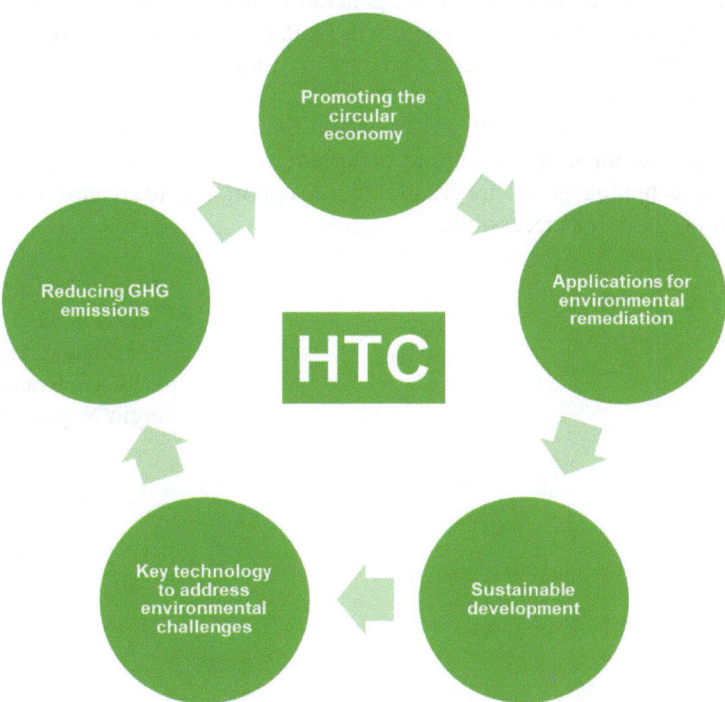

As far as the authors are aware, numerous texts on HTC can be found in the literature, but they are often adapted in such a way that they are only accessible to a specific audience. These texts use highly specialized terminology and deal with topics that can be complex, making it difficult to capture the interest of those who wish to start studying this technique.

For this reason, this book is intended to reach students or junior researchers who aspire to learn about the fundamentals of the technique in an accessible and understandable way, without neglecting the precision and scientific depth needed to establish a solid base of knowledge.

Throughout the chapters, we will explore the meaning of hydrothermal carbonization, going through its basic principles, the main chemical reactions involved, the types of waste used, practical considerations, industrial aspects and the main challenges and current innovations, all with special emphasis on a practical and simple way, supported by problems and questions that invite reflection and allow students to apply what they have learnt in an active way.

The team we form is sensitively involved with sustainability and efficient resource management and we understand that these are critical challenges, which were already a problem in the past, present and future. We are therefore committed to hydrothermal carbonization as a promising alternative for the valorization of waste, contributing to the reduction of carbon footprint and the generation of useful products from materials that might otherwise be considered waste. By training a new generation of students in this technology, we aim to encourage greater innovation and adoption of sustainable solutions in the future.

We hope that this book will serve not only as a source of knowledge but also as an inspiration for those seeking to contribute to the development of clean and sustainable technologies. We are grateful for the opportunity to create a resource that is both informative and accessible.

Badajoz, Spain

Beatriz Ledesma Cano
María Alonso Sánchez
José Manuel Díaz Rasero
Silvia Román Suero
Sergio Nogales Delgado

Contents

Chapter 1
Overall Picture: The Role of Sustainable Processes and Renewable Resources

1.1 Sustainable Development

Our globalized society faces new and complex challenges, many of them having to do with the environment and depletion of resources. On the one hand, the increasing global population demands more and more energy and products, with the subsequent environmental impact related to the manufacture of these goods, but also including the exploitation of raw materials and their final use or management. On the other hand, the so-called consequences of exploiting petroleum resources are especially highlighted nowadays, with a special attention to different factors related to climate change such as greenhouse gas emissions or water/soil pollution. Additionally, oil and natural gas dependence in some continents and regions has derived in the increase in dependence relations and inequality between developed countries and developing ones. This is not a trivial issue, as the role of non-renewable energy has been historically used as a geopolitical asset for decades to gain international relevance. Unfortunately, different conflicts are currently active (with negative expectations in the long term), where different geopolitical measures directly affect to different supplies and prices (most of them basic needs), normally at the expense of social and economic well-being of society in general.

Under these circumstances, it is no wonder that a global concern about the environment has arisen in society, pointing out the interest (at least seemingly) of different governments or international institutions in green technologies in order to contribute to the sustainable growth of both developed regions and, especially, developing areas whose role in this aspect will be crucial. As a consequence, there are plenty of national and international guidelines or standards that foster green practices at different levels, from households to industries.

A paradigmatic example is the sustainable development goals (SDGs) [1], apparently adopted by all United Nations member states, which cover different aspects of sustainability from local to global like the following: the end of poverty and hunger; the promotion of healthy habits and well-being; fostering education and

B. Ledesma Cano et al., *Introduction to Hydrocarbonization*,
SpringerBriefs in Applied Sciences and Technology,
https://doi.org/10.1007/978-3-031-70039-2_1

equality (especially related to gender); soil, water and air preservation; improvement of working conditions and technology; the implementation of sustainable cities and communities; responsible consumption and production, etc. In that sense, the role of partnership and institutions is vital, and scientific community will have its say, contributing to the progress in the implementation of green technologies with circular economy.

In this way, there are different sustainable processes focused on the production of energy and different bioproducts that can respond to the abovementioned emerging challenges. For this purpose, these technologies usually share some points in common, like the following:

- These processes present a high atom economy, where most of the products (some of them intermediate and other by-products) are used for different purposes, not being released to the environment, and considerably reducing the subsequent environmental impact.
- They are usually based on natural resources, many of which are wastes produced in industrial or agro-industrial processes, implying a double advantage. On the one hand, wastes with a challenging management can be valorized. On the other hand, the use of natural raw materials for these processes derives in the production of biodegradable products, with lower environmental impact compared to other products based on petroleum.
- Normally, sustainable technologies are quite adaptable to different contexts, as different raw materials can be used. In this sense, this fact could contribute to the sustainable development of different countries or regions, where different resources (including wastes) could be reused in sustainable processes, reducing the external energy (and material) dependency, and consequently improving the standard of living of general population.
- Additionally, some techniques are versatile, obtaining outcomes for different purposes. For instance, some processes can be used for energy purposes as well as the production of different chemical compounds or materials that are equally interesting for industry.

Considering the above, many different technologies could be within this framework, offering a wide range of alternatives and possibilities to contribute to the sustainable development and the implementation of green technologies on a global scale. Specifically, hydrothermal carbonization (HTC), as explained in further subsections and chapters, has been described as a promising alternative for some technologies based on traditional and polluting processes.

1.2   Introduction to Hydrothermal Carbonization

Hydrothermal carbonization or hydrocarbonization (HTC) is a novel and promising process for the conversion of various biomass and other non-biomass materials or wastes such as sludge. It was originally discovered by Friedrich Bergius (a German

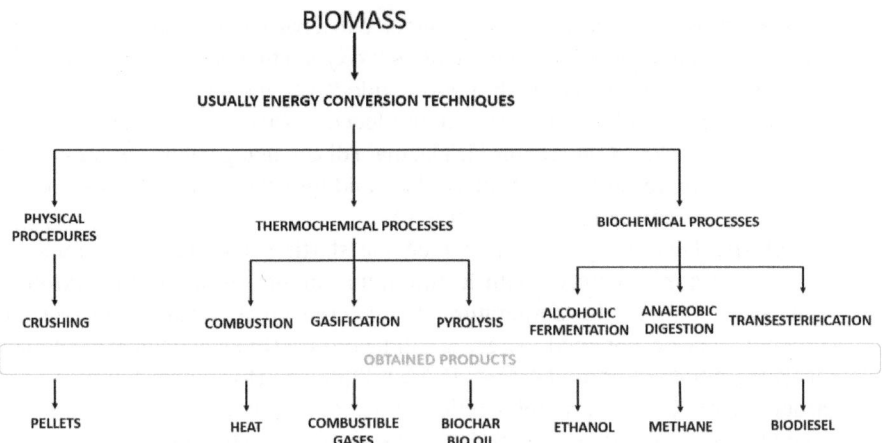

Fig. 1.1 Techniques for energy conversion of biomass

chemist and industrialist), who was awarded the Chemistry Nobel Prize in 1913, recognition to his research on how high-pressure water treatment can transform glucose. However, it is in recent times, during the last decade, that this technique has achieved a great development.

To understand the relevance of HTC treatment and its advantages over other energy recovery or carbon production techniques, it is necessary to compare it with other processes generally used for these purposes. Thus, within the field of biomass, the most traditionally used techniques (Fig. 1.1) are the following:

- **Physical procedures**: These processes basically consist of crushing and reducing the size of the biomass particles to obtain pellets or briquettes. The main disadvantage of these processes is that in order to manufacture these pellets or briquettes, a previous biomass drying treatment is necessary, which entails a high previous energy cost. Also, pelletizing of crushed biomass requires not only heat and pressure but also chemical binders.
- **Thermochemical processes**: These processes include combustion, gasification, torrefaction and pyrolysis.

  - **Combustion**: It consists of chemical reactions in which the fuel (in this case a biomass or waste) reacts with oxygen (comburent). As it is an exothermic reaction, it produces thermal energy (heat). Subsequently, this heat can be directly used or be linked to the generation of electricity in steam power plants. The main disadvantage of this process is that the original fuel must be as free of moisture as possible, in addition to greenhouse gas emissions such as CO_2, NO_x or CO. In addition, the presence of specific elements and the amount of composition of ash narrows the range of good candidates as fuels, in the field of biomass wastes.

- **Gasification**: In this case, a fuel is subjected to a partial or incomplete combustion, i.e. a combustion that occurs with less oxygen than necessary. As a result, combustible gases such as carbon monoxide (CO), methane (CH_4), hydrogen (H_2) and other combustible gases are produced. To achieve this, high temperatures of up to 1200 °C are required. The main disadvantage of this process is the high temperatures at which it takes place and thus the energy costs involved, as well as the emission of pollutant gases.
- **Pyrolysis**: This is a process in which the starting material is decomposed in an inert atmosphere or with a minimum amount of air by the action of heat. Depending on the temperature at which this process takes place, various products can be obtained. Thus, a solid product (biochar) can be obtained at temperatures of around 400–500 °C, a liquid oil (for higher temperatures, around 700 °C) and a combustible gas (at temperatures of 900–1000 °C). The immediate disadvantage of these processes is also the high energy cost of reaching these temperatures as well as the potential toxicity of the oil obtained.

- **Biochemical processes**: These processes are based on the action of microorganisms that decompose the biomass to obtain various compounds. They are complex processes from a chemical point of view. Depending on the type of specific decomposition reaction and the type of starting biomass, three main processes can be classified according to the different products obtained, namely: alcoholic fermentation processes by means of which ethanol is obtained; anaerobic digestion processes, that yield methane, and transesterification processes, from which biodiesel (fatty acid methyl or ethyl esters) can be obtained.

From the analysis of energy needs associated to the former processes, a number of conclusions can be drawn. Firstly, most of the abovementioned techniques require high energy costs due to the high temperatures involved. In addition, previous cost-intensive drying of the biomass for its subsequent transformation is required. The second conclusion is greenhouse gas emissions. Some techniques (combustion, pyrolysis or gasification) have the problem of emitting polluting and greenhouse gases, depending on the waste or biomass in question. In addition, some of them such as pyrolysis produce a highly polluting final oil or tar that is difficult to use and eliminate.

Regarding techniques such as anaerobic digestion or alcoholic fermentation, their main problem is the long time needed to obtain the final product, either ethanol or methane. In addition, large quantities of biomass or waste are needed for these processes to work.

Thus, according to the above, the hydrothermal carbonization process (HTC) is presented as an ideal alternative to all the processes previously described.

1.2.1   What is HTC?

The HTC technique has gained a lot of importance in the last decade due to the multiple performance advantages and the potential and varied applications of the final products obtained.

The HTC process is a thermochemical biomass conversion technique. It basically consists of the degradation of its macroconstituents (cellulose, hemicellulose and lignin) in the presence of water (more information about this in Chap. 2). This reaction requires low temperatures compared to other processes, around 180–250 °C and high pressure (autogenerated by the system) in order to obtain subcritical water, as observed in Fig. 1.2. In this phase diagram, there are the typical areas where different states can be found, and two interesting points: the triple point and the critical point. The former is the point (with a certain pressure P_{tp} and temperature T_{tp}) at which the solid, liquid and vapour states are in equilibrium. Regarding the critical point, it is the point (with its corresponding critical pressure, P_{cr}, and critical temperature, T_{cr}) at which liquid and gas phases are indistinguishable due to their common properties, like density. Table 1.1 shows the main values for these points in the case of water, the main element used in HTC. Thus, the temperature used in HTC is usually below the critical temperature of water, whereas pressure range includes the critical pressure of water. In other words, HTC works in subcritical conditions, according to Fig. 1.2. This whole process is carried out in a closed reactor where the biomass is mixed with water.

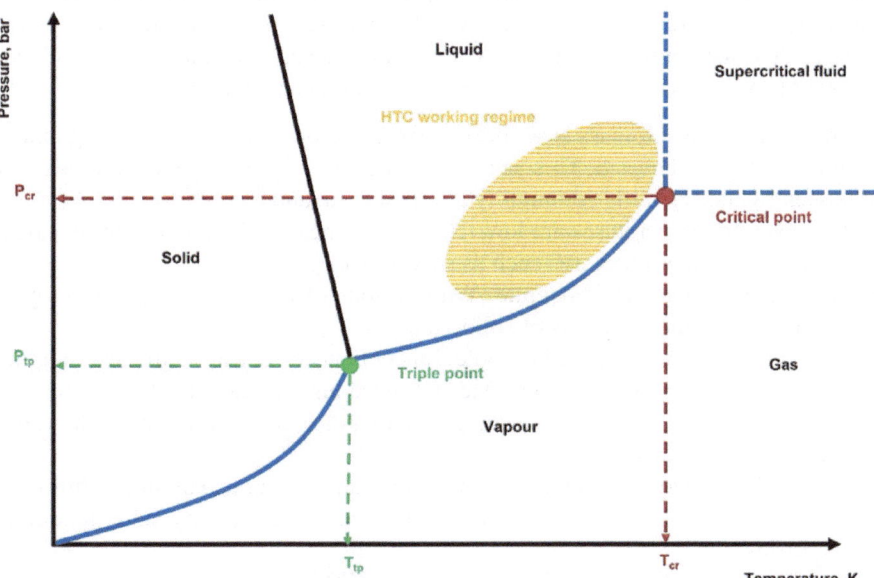

Fig. 1.2 Phase diagram of water and working conditions for HTC process

Table 1.1 Most significant points in the phase diagram of water

Point in phase diagram	Pressure, bar	Temperature, °C
Triple point, TP	$P_{tp} = 0.0061$	$T_{tp} = 0.01$
Critical point, CR	$P_{cr} = 220.64$	$T_{cr} = 373.95$

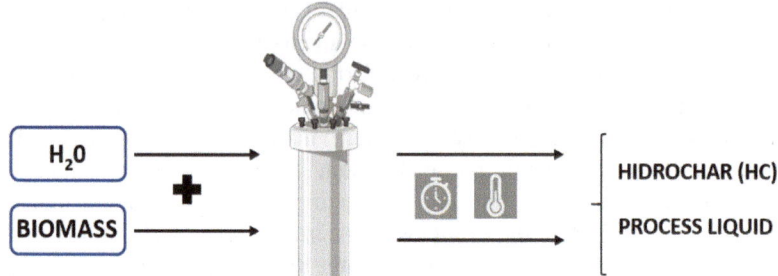

Fig. 1.3 HTC process

The final products obtained are a liquid phase (normally called process liquid) and a solid phase called hydrochar or HC (see Fig. 1.3).

Small quantities of gases such as CO_2, CH_4, H_2 and others are also produced, but in practically negligible quantities. The amount produced of gases, solid and liquid, mainly depends on time, temperature and the biomass-water ratio, among other variables that have been studied to a lesser extent (addition of catalysts, void reactor volume, particle size, etc., that will be analysed in more detail in Chap. 4).

With this first approach to the HTC technique, several advantages of HTC over the other processes described in the previous section can be described:

- Pre-drying of the biomass is not necessary. Therefore, biomasses with high moisture content can be used. This would save the cost of pre-drying the biomass and is seen by the scientific community as a major advantage.
- The reaction temperatures are low. This would mean a significant energy saving compared to other processes such as pyrolysis or gasification.
- The reaction times are moderate. With periods of time of around 1–3 h, good results are obtained in terms of solid yield.
- The operation of the reactor is simple. It is only necessary to set the reaction time and temperature and the process does not need any additional control other than providing a proper maintenance that guarantees the good functioning of the valves and security devices.
- The solid obtained (hydrochar) and the liquid can be separated by simple filtration.
- Solid product (hydrochar) presents improved dewaterability and pelletizing properties.
- Non-biomass materials can be used. In fact, HTC has proven to be a promising disposal technique for all kinds of wastes, including sewage sludge [2] or livestock effluents.

1.3 Relevance and Applications of HTC Products

1.3.1 Hydrochars

Hydrochar (HC) is the final solid product that is obtained at the end of the HTC process. This solid, in general, has a greater high heating value (HHV) than the starting biomass. It is an interesting starting point, as these HCs could be used as the basis for pellet manufacturing, being employed in pellet stoves or small-scale steam plants located in rural areas. This way, two basic objectives, can be achieved: the reduction of evolved gas (most of them polluting) upon combustion and the development of rural areas, as the energy requirements of these areas could be fulfilled with an affordable and cheap material.

Combustion of HCs could be, however, environmentally harmful due to the presence of different elements like nitrogen, chlorine or sulphur. As a matter of fact, an innovative and interesting research field is based on monitoring different compounds based on specific elements (such as N, P, S or Mg) that evolved during HTC reaction.

The different reaction conditions (mainly reaction time and temperature) and the type of the biomass used, condition the extent of the migration of interesting elements to the solid or its stability and form in the processing water. In this sense, some studies [3] have established the procedure to carry out the monitoring of N during the process. This point is very interesting, as the potential applications of HCs can depend on the presence or absence of N. Thus, these products can be used for electrode manufacture for batteries [4] or fertilizer (rich in N) production [5]. These are two examples where the presence of nitrogen in HCs is beneficial. However, if HCs were used for combustion (in the form of pellets), as previously explained, the aim would be to limit the presence of N in these products, to avoid further NO_x emissions.

Along with the relevance of nitrogen content in HCs, another important use of these solids is as adsorbents [6]. Thus, due to the high specific surface of this solid, along with the possible presence of some functional groups in its surface (such as phenolic, phosphoric, amine, carboxylic or peroxide, among others), this product could be an effective adsorbent of specific compounds (depending on their acidity, polarity, etc). Adsorbent materials have a wide range of uses. Among them, these materials can be used to develop filters for water purification [7], presenting other different applications like the manufacture of solar panels or other electronic devices. The porosity (volume and pore size distribution) that a feedstock can yield after carbonization or activation depends on the technique used. In that sense, this is another interesting research field related to the solid products obtained in HTC since, as a rule, carbons obtained through HTC have less porosity than those obtained through pyrolysis [8]. Therefore, achieving a suitable activation of HCs is especially interesting. It should be noted that, depending on the starting biomass, this activation could be more or less difficult, including heating rate, temperature, activating agent type and flow and others.

Recent research works are focused on the final state of HCs, studying the possible influence of temperature or nitrogen level on their final characteristics. Thus, novel research works have pointed out the luminescence of some secondary HCs (which will be analysed in Chap. 2).

A very important area that would allow a more efficient control of this process (for its possible use at industrial level) is the determination and understanding of the reaction mechanisms taking place in this process (see Chap. 2). Thus, the control of the adequate variables to obtain a certain quantity of HCs with specific characteristics and the development of models could imply an important advance when it comes to the practical implementation of this process at industrial level.

1.3.2 Liquid Products or Water Used in HTC

The liquid resulting in HTC processes is also an interesting product from a research perspective. This liquid, often called water process, presents different characteristics depending on the kind of starting biomass and the reaction conditions. Even though there is a wide range of compounds that can be found in HTC liquid, they can be classified into different groups according to their common nature, like the following: extractives, sugars (sucrose, glucose or fructose), phenolic compounds (like phenol or catechol), furfural and its derivatives, other organic acids (such as formic, lactic or acetic acids, among others) and inorganic acids. Consequently, the purification of these compounds could be an interesting point for the valorization of this liquid. Thus, among the numerous compounds and elements included in this liquid, it should be noted the presence of several products due to their potential interest in chemical industry, like 5-hydroxymethylfurfural [9] or levulinic acid [10]. The former could be an interesting starting point for the production of interesting compounds such as 2,5-furandicarboxylic acid for polyester production, 2,5-dimethylfuran (used as biofuel) or 2,5-Bis(hydroxymethyl)furan (which could be used as a bio feedstock). Regarding levulinic acid, it presents a wide range of possibilities, as it can take part in the production of pharmaceutical products, plasticizers, additives or biodegradable herbicides. Thus, as explained in further chapters, the direct use of this product, as well as some included compounds, could make the liquid product obtained during HTC an interesting component with multiple applications.

Apart from these compounds, the presence of interesting elements such as P, N or K could be decisive for their final usage. These elements can be obtained from HTC liquid to obtain interesting fertilizers for agricultural purposes. Some studies [2, 7, 11] have found the contrary effect, that is, these works have proved the herbicide effect of this liquid by-product obtained from sewage sludge in wastewater treatments. Thus, the use as fertilizer or herbicide could depend on the presence of some substances in this liquid due to the starting biomass. In the abovementioned study [11], the herbicide effect of HTC liquid obtained from a wastewater plant was assessed, whereas its use as fertilizer when this by-product was obtained from digestate derived from cow manure was studied in other works [12]. Nevertheless, the herbicide or fertilizer effect of this liquid for the same biomass could depend on its concentration.

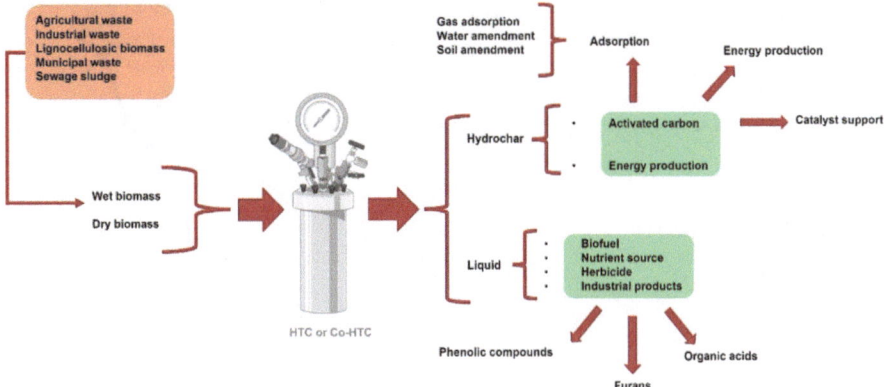

Fig. 1.4 HTC in a biorefinery context

To sum up, and as abovementioned, HTC could be an interesting component in a biorefinery context, taking part in more complex processes in order to valorize some wastes. Thus, as observed in Fig. 1.4, there are endless opportunities for this process, especially considering the wide range of sources that can be used and the possibilities for the main products.

As observed, a wide range of biomass from different sources can be used in HTC, with the possibility of using wet biomass (including water) or dry biomass, which should be combined with water or another wet biomass. In that sense, as explained in further chapters, the use of wet biomass could be suitable for this purpose, and its combination with dry biomass, which is called co-HTC, could be another alternative for the valorization of dry biomass in this process, avoiding the addition of water and reducing costs. On the other hand, this figure shows the multiple alternatives for the different products obtained in HTC (or co-HTC), with multiple alternatives that can be interesting for pharmaceutical industry, agriculture or environmental technology, among other specific uses.

1.4 Scientific Interest

As a consequence of the above, the scientific interest in HTC has gained importance, especially in the last 10–15 years, proving the novelty of this technique in the current scientific and technological scenario. As observed in Fig. 1.5, where the number of publications per year is included, there has been a considerable increase in publications (mainly books, book chapters or scientific papers) since 2008, currently obtaining a publication rate of at least 900 publications per year, observing an exponential growth. This fact indicates the promising future of this field, with the subsequent possibility of adapting this technology at industrial scale in the long run.

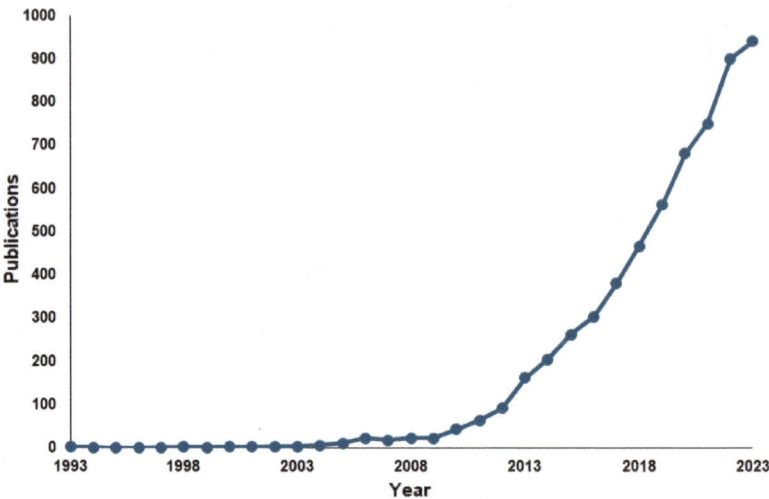

Fig. 1.5 Evolution of the number of publications about HTC. *Source* Scopus [13]

If these articles are sorted by country, interesting data can be obtained, as observed in Fig. 1.6. Thus, and even though China contributes with the majority of publications, there is a uniform distribution of publications all over the world, with the USA, Germany, India and Spain ranking the first places. Another interesting point is the considerable participation of countries from the five continents, which could be an interesting starting point for the sustainable implementation of this technology, especially in developing countries or regions. Indeed, there are already industrial HTC in Europe, Asia and America. As explained in further chapters, this technology is versatile and easily adaptable to different resources. That is the reason why research distribution is homogeneous all over the world, as HTC could be a very interesting starting point for the implementation of green technologies, especially in developing countries, where different wastes (some of them typical of each region) could be suitable for their use through this technology.

It should be noted that, even though it is a relatively novel technology compared to other thermochemical processes, there has been a considerable impact of this process in the last ten years, as it can be observed in Fig. 1.7. In that sense, the publications related to HTC increased close to 600%, whereas the rest of the abovementioned processes present lower (although not negligible) increases, around 200%, which is equally interesting, considering the technological maturity of these processes. This fact proves the promising possibilities of HCT from a scientific and technical perspective, implying another alternative, or even a complementary technology, for different wastes or sources.

Regarding these publications, their main keywords and their interrelations are included in Fig. 1.8. This could give us an interesting research scenario, as most of them are multidisciplinary works related to other fields. As observed, there are different clusters (that is, different keywords with a close connection, represented

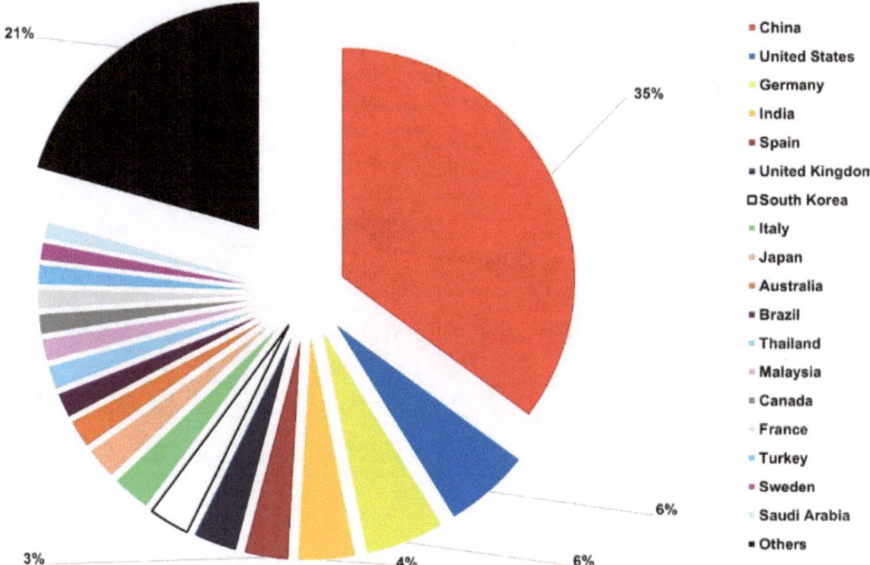

Fig. 1.6 Publications about HTC sorted by country. *Source* Scopus [13]

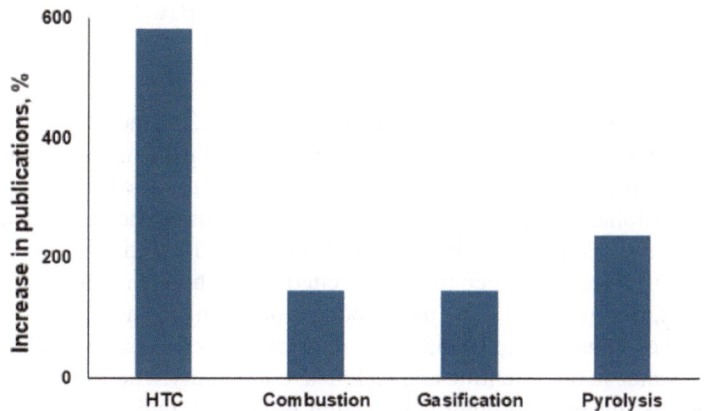

Fig. 1.7 Increase in publication about different subjects in the last ten years. *Source* Scopus [13]

with the same colour) focused on different topics, most of them covered in this book. This way, one of the main products obtained during HTC, that is, biochar, is commonly related to their use as activated carbon for adsorption for different purposes (including nutrient recovery in soils or environmental remediation), presenting interesting characteristics such as high porosity, which should be characterized. Also, the liquid obtained at the end of the process is equally important, with some interesting

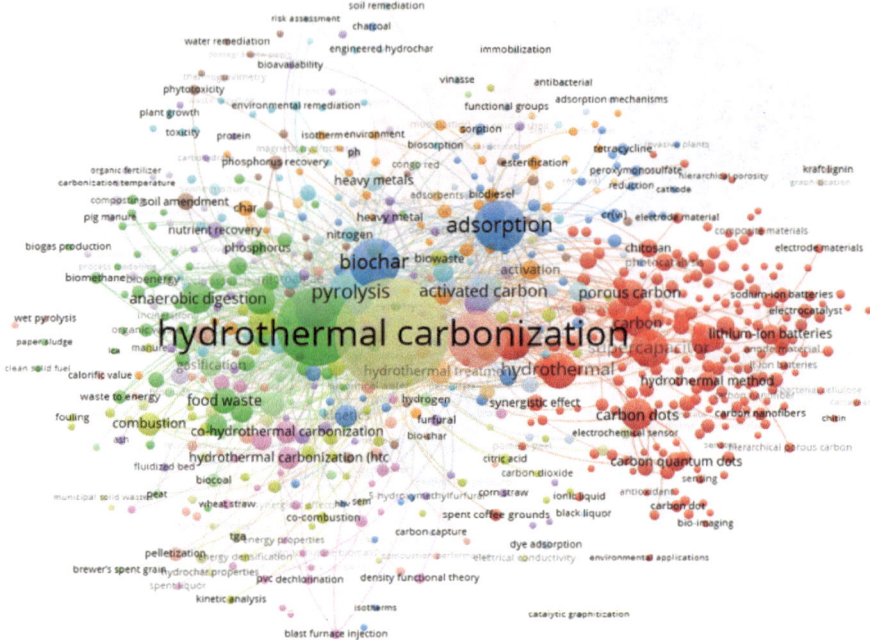

Fig. 1.8 Interrelation of main keywords in publications related to HTC. Figure obtained by using VOSviewer. *Source* Scopus [13]

properties that could make it suitable for some processes. On the other hand, the co-hydrothermal carbonization, where the combination of different wastes can present very important results, plays another important role, with different wastes such as food waste combined with traditional agro-industrial wastes such as brewer's spent grain. In any case, concepts included in this figure (such as adsorption, functional groups, porosity, etc.) will be continuously cited in this book in many ways.

To sum up, in this chapter, the main foundations of hydrothermal carbonization have been covered, including the main possibilities of this technology and the current interest and status in the scientific and technical community. The following chapters will be focused on the thorough explanation of these foundations, including the possible technical implementation at industrial scale.

1.5   Questions

1. HTC was discovered by:

   (a) Franz Fischer
   (b) Friedrich Bergius

    (c) Hans Tropsch

    (d) Wilhelm Ostwald

2. What are the main reagents present in HTC?

    (a) Nickel nitrate hexahydrate and water

    (b) Water and biomass

    (c) Methanol and vegetable oil

    (d) Biomass and sodium nitrate

3. Which of the following is true about HTC?

    (a) It is carried out in supercritical conditions

    (b) Its temperature range is usually below 373.95 °C

    (c) Pressure can never exceed the critical pressure of water

    (d) Options a and b are correct

4. What are the main products obtained in HTC?

    (a) Carbon dioxide and charcoal

    (b) Hydrochar and HTC liquid

    (c) Ammonia and hydrogen sulphide

    (d) Methane and fatty acid methyl esters

5. Which of the following are characteristics of HTC to be considered a green technology?

    (a) Possibility of waste valorization

    (b) Versatility, as many sources can be used in HTC

    (c) Variety, as different products in solid and liquid state can be obtained

    (d) All of the above

6. Which of the following about HTC is wrong?

    (a) Its scientific interest has arisen in the last fifteen years

    (b) The USA is the main country devoted to the scientific research in this subject

    (c) Not only wet biomass can be used in HTC

    (d) None of the above

Competing Interests The authors have no conflicts of interest to declare that are relevant to the content of this chapter.

References

1. United Nations (2015) Transforming Our World: the 2030 Agenda for Sustainable Development. https://www.un.org/sustainabledevelopment/sustainable-development-goals/. Accessed 25 Apr 2024
2. K. Czerwińska, A. Marszałek, E. Kudlek, M. Śliz, M. Dudziak, M. Wilk, The treatment of post-processing liquid from the hydrothermal carbonization of sewage sludge. Sci. Total. Environ. **885**, 163858 (2023). https://doi.org/10.1016/j.scitotenv.2023.163858
3. A. Kruse, F. Koch, K. Stelzl, D. Wüst, M. Zeller, Fate of nitrogen during hydrothermal carbonization. Energy Fuels **30**(10), 8037–8042 (2016). https://doi.org/10.1021/acs.energy fuels.6b01312
4. J. Tan, H. Chen, Y. Gao, H. Li, Nitrogen-doped porous carbon derived from citric acid and urea with outstanding supercapacitance performance. Electrochim. Acta **178**, 144–152 (2015). https://doi.org/10.1016/j.electacta.2015.08.008
5. F. Fornes, R.M. Belda, A. Lidón, Analysis of two biochars and one hydrochar from different feedstock: focus set on environmental, nutritional and horticultural considerations. J. Clean. Prod. **86**, 40–48 (2015). https://doi.org/10.1016/j.jclepro.2014.08.057
6. R. Sivaranjanee, P.S. Kumar, G. Rangasamy, A recent advancement on hydrothermal carbonization of biomass to produce hydrochar for pollution control. Carbon Lett. **33**(7), 1909–1933 (2023). https://doi.org/10.1007/s42823-023-00576-2
7. S. Román Suero, B. Ledesma, A. Álvarez, C. Herdes, Towards sustainable micro-pollutants' removal from wastewaters: caffeine solubility, self-diffusion and adsorption studies from aqueous solutions into hydrochars. Molecular Phys. **116** (2018). https://doi.org/10.1080/002 68976.2018.1487597
8. H.S. Kambo, A. Dutta, A comparative review of biochar and hydrochar in terms of production, physico-chemical properties and applications. Renew. Sustain. Energy Rev. **45**, 359–378 (2015). https://doi.org/10.1016/j.rser.2015.01.050
9. F. Menegazzo, E. Ghedini, M. Signoretto, 5-Hydroxymethylfurfural (HMF) production from real biomasses. Molecules (Basel, Switzerland) **23**(9), 2201 (2018). https://doi.org/10.3390/molecules2309220
10. J. Chaparro-Garnica, M. Guiton, D. Salinas-Torres, E. Morallon, E. Benetto, D. Czorla-Amorós, Life Cycle assessment of biorefinery technology producing activated carbon and levulinic acid. J. Clean. Prod. **380**, 135098 (2022)
11. K. Czerwińska, A. Marszałek, E. Kudlek, M. Śliz, M. Dudziak, M. Wilk. The treatment of post-processing liquid from the hydrothermal carbonization of sewage sludge. Sci. Total Environ. **885**, 163858. https://doi.org/10.1016/j.scitotenv.2023.163858
12. S. Celletti, M. Lanz, A. Bergamo, V. Benedetti, D. Basso, M. Baratieri, S. Cesco, T. Mimmo, Evaluating the aqueous phase from hydrothermal carbonization of cow manure digestate as possible fertilizer solution for plant growth. Front. Plant Sci. **12**, 687434 (2021). https://doi.org/10.3389/fpls.2021.687434
13. Scopus Scopus. https://www.scopus.com/home.uri. Accessed Feb 2024

Chapter 2
Basic Principles of HTC

2.1 Introduction

Hydrothermal carbonization technique (HTC) is an innovative process based on the fundamental principles of moderate high pressure and temperature, together with the crucial presence of water as a reagent and solvent. This technique offers a promising route for the conversion of biomass and organic waste into valuable products, with potential applications in various industries, from biofuel production to waste treatment and soil improvement [1].

In HTC process, organic matter is subjected to extreme conditions, with temperatures typically in the range of 180–260 °C and pressures above atmospheric (slightly higher than the corresponding saturation temperature, although other factors like the reactor void volume have to be also considered). These conditions facilitate a series of complex chemical reactions, including hydrolysis, dehydration, decarboxylation, aromatization and recondensation. These reactions lead to the decomposition of organic matter and the formation of new compounds, resulting in the generation of hydrochar, a carbon-rich solid material with a porous structure.

In addition to the solid hydrochar, a liquid phase known as process water (PW) is obtained, and a gaseous phase containing volatile compounds and gases such as CO_2, CH_4, and H_2. Precise control of the reaction conditions, including temperature, pressure, water/biomass ratio and residence time, is essential to optimize the yield and quality of the products obtained by HTC. Understanding the reaction mechanisms involved in this process is essential to maximize its efficiency and feasibility in various industrial and environmental applications.

In summary, HTC is a promising technique that exploits the basic principles of high pressure, temperature, and the presence of water to convert biomass and organic waste into valuable products, with the potential to contribute significantly to the transition towards a more sustainable and renewable resource-based economy.

B. Ledesma Cano et al., *Introduction to Hydrocarbonization*,
SpringerBriefs in Applied Sciences and Technology,
https://doi.org/10.1007/978-3-031-70039-2_2

2.2 Chemical Reactions Involved in HTC

Hydrothermal carbonization (HTC) is a complex and dynamic process that involves, under high pressure and temperature conditions, organic matter in the presence of water undergoing a series of significant chemical reactions such as hydrolysis, dehydration, decarboxylation, aromatization and recondensation, resulting in the generation of a carbon-rich solid material with a porous structure that is formed from the biomass or organic waste during the process, with high energy density and added value, called hydrochar [1].

In addition, a liquid phase or process water (PW) consisting of water-soluble products and containing a variety of organic compounds such as carboxylic acids, alcohols, aldehydes and ketones, and inorganic compounds, and a gas phase containing volatile compounds and gases such as CO_2, CH_4, H_2 and volatile organic compounds are also obtained as end products.

$$Biomass + H_2O \rightarrow Hydrochar + Liquid + Gas$$

When the HTC reaction is initiated, there are intermediate products formed in the reaction medium that are not directly observable at the end of the reaction but influence its development. These intermediate products are chemical species that are consumed in stages prior to the final stage and end up being transient. Examples include reactive molecules, ions, free radicals or coordination complexes, among others. These intermediates can influence different process variables, such as the speed, selectivity or efficiency of the HTC.

A clear example of this final product is called secondary hydrochar (SHC), which is produced due to the liquid composition of the medium and generates another solid that remains in the dispersion of the medium, is a result of recombination of macromolecules that aggregate to form nanoparticles. Sometimes, it is not identifiable to the naked eye, but it does influence the stability of the final products.

However, the crucial component for HTC to occur is water because it acts as a reaction medium, assisting in the conversion of biomass and carbonaceous products, since water is present in the organic materials of these products.

In HTC, water is a precursor to the generation of molecules that then undergo condensation and polymerization reactions for the formation of final carbonaceous products.

In the process of hydrolysis at temperatures and pressure that can attack the chemical bonds present in organic materials, there is a splitting into hydroxyl ions (OH^-) and protons (H^+). These molecules are usually used for polymerization and condensation reactions. In the dehydration step, recombination of these hydroxyl bonds with adjacent C atoms occurs, resulting in a product with more complex carbon structures [2].

As for the successive stages, in dehydration, carbohydrates and other organic compounds lose water molecules remove hydroxyl groups, due to the high temperature and pressure conditions, which significantly carbonizes the biomass by reducing the oxygen-to-carbon (O/C) and hydrogen-to-carbon (H/C) molar ratios [3].

The glucose dehydration process is depicted below:

$$C_6H_{12}O_6 \rightarrow 6C + 6H_2O$$

During polymerization, organic monomers join to form larger and more complex polymers. This stage leads to the formation of larger and more stable carbon structures. An example is the polymerization of glucose to form cellulose:

$$nC_6H_{12}O_6 \rightarrow (C_6H_{10}O_5)n + nH_2O$$

In the condensation process, there is the joining of different organic molecules to also generate larger and more complex compounds, such as C–C bonds. These bonds influence the resulting products, increasing their density and stability. For example, this reaction occurs in the condensation of a compound with two phenolic molecules (C_6H_6O), when they react with oxygen to form a new condensed product $(C_{12}H_{10}O_4)$ obtaining water as a by-product.

$$2C_6H_6O + O_2 \rightarrow C_{12}H_{10}O_4 + 2H_2O$$

In the C–C bond formation reaction, three-dimensional structures that are more stable than the previous ones are created. An example would be the polymerization of a C–C bond where two ethylene molecules (C_2H_4) recombine to form butane (C_4H_8).

$$C_2H_4 + C_2H_4 \rightarrow C_4H_8$$

During decarboxylation and decarbonization, an amount of gas (mainly composed of CO_2 and CO) is generated by partially degrading carboxyl and carbonyl groups, also contributing to carbonization, but to a lesser extent during HTC [4].

Aromatization involves the conversion of linear organic compounds into aromatic compounds by rearrangement of chemical bonds. The carbon content in the aromatic rings results in increased stability and resistance to microbial degradation.

Although the processes occurring within HTC are known, these different reactions can occur in parallel and influence each other. For example, in the formation of carbonaceous products, the dehydration reaction usually takes place first, followed by polymerization and condensation, but sometimes the latter two are formed before the dehydration process is completed. Reversely, some degradation compounds can undergo hydration reactions, as it is the case with 5-hydroxymethylfurfural, that is formed by dehydration of glucose, but then can be transformed by hydration to levulinic acid [5]. A small diagram of the reactions associated to biomass HTC is shown in Fig. 2.1.

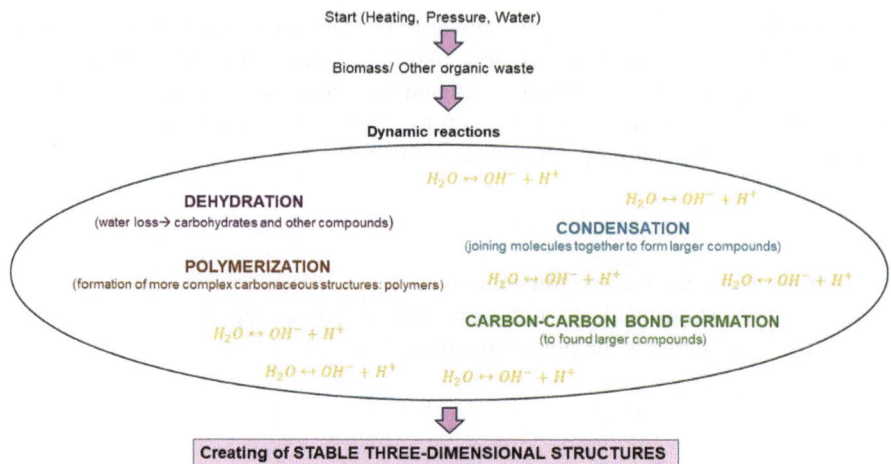

Fig. 2.1 Techniques for energy conversion of biomass

It is crucial to understand these fundamental reactions in order to optimize the HTC process and obtain the desired products with maximum efficiency.

2.3 Catalysis and Catalysts Used in the Process

The existence of so many compounds on HTC system (biomass, organic or inorganic degradation products, etc.), poses a great challenge because each of them has a different decomposition temperature that makes them react or not, depending on many variable factors. Adding catalysts into HTC can improve the process, optimize the obtaining of products and/or improve the economical balance.

Therefore, depending on the choice of the catalyst used, aspects of the process such as the reaction rate, the selectivity of the final products, as well as the stability of the intermediate products will vary significantly. In conclusion, catalysts are introduced in HTC to try to improve both the properties of the final product and the process itself [6].

2.3.1 Catalyst Selection

The appropriate choice of catalyst is a crucial aspect in HTC and depends on several factors such as the composition of the organic substrate, the reaction conditions and the desired properties of the final products.

The feedstock used in HTC can vary widely in nature and chemical composition. For example, lignocellulosic materials like biomass, a common substrate in HTC, are mainly composed of cellulose, hemicellulose and lignin, each with different functional groups such as alcohols, aldehydes, carboxylic acids and phenols. The presence of these functional groups may require the selection of specific catalysts that selectively interact with them to promote the desired reactions.

Reaction parameters such as temperature, pressure, residence time and pH of the medium can also influence the efficiency and selectivity of the catalyst. Some catalysts may perform better under certain reaction conditions, whereas others may be more versatile and adapt to a wider range of conditions.

The residence time, i.e. the time that the catalyst is in contact with the medium, is also very influential in the formation of final products. Some catalysts need longer residence times to obtain a conversion of the reactants, while others can show high activity with a short residence time.

The pH of the medium, both at the start and during the reaction, has a significant impact on the activity and selectivity of the catalyst, especially at the extremes, if the medium has a particularly acidic or basic pH.

2.3.2   Types of Catalysts Used

In the HTC process, a variety of catalysts has been used to facilitate and control the chemical reactions involved and can be classified into two main categories: organic and inorganic catalysts (see Table 2.1). Classification of catalyst can be made based on various factors like composition, associated retention time and the temperature, as it is described next [6]:

Table 2.1 Classification of catalysts used in the HTC process

Catalysts	
Organic catalysts	Inorganic catalysts
Organic acids	Strong mineral acids
Alcohols	Strong mineral bases
Other organic catalysts	Sulphate, persulphate and sulphite salts
	Nitrate salts
	Metal oxides
	Hydrogen peroxide
	Strong mineral acids
	Metal chlorides
	Other inorganic catalysts

Organic Catalysts

- **Organics acids**

During HTC, there is a reaction in which organic acids (such as acetic, formic or citric acid) make the medium become an acid solvent favouring hydrolysis, dehydration and carbonization reactions that are significantly influenced by temperature. The presence of these compounds on the process water (type and concentration) affects the extent of their potential adsorption on the Hydrochar (HC) surface and will have effects on the structural (micro and mesoporosity) and chemical (functional groups, microspheres...) features.

- **Alcohols**

These are protic solvents, i.e. they are polar compounds with dissociable water atoms in their molecular structure. This release of hydrogen could lead to a dehydration reaction in the HTC, which could result in a repolymerization leading to greater HC yield. They accelerate the feedstock decomposition reactions and reaction rate. Examples of alcohols that can be used as catalysts in HTC are methanol, ethanol or isopropanol, among others.

- **Other organic catalysts**

These catalysts enhance the efficiency of HTC and the quality of HCs. This group includes calcium salts, water-soluble polymers and organic compounds such as formaldehyde.

Calcium salts with compounds such as calcium acetate or calcium lactate show HC densification properties thus improving its fuel quality. As an example, cationic polyacrylamide can help in dehydration by removing sulphur [7]. Finally, the use of formaldehyde in HTC promotes polymerization, which leads to higher reaction efficiency, better HC quality with improved calorific value [8].

Inorganic Catalysts

The addition of such acids to the medium aids the formation of compounds useful for improving HTC or HC properties. The following are some of the acid catalysts used in the process:

- **Strong mineral acids**

These acids can increase the stability and energy value of the HCs. Some of the most common acids are hydrochloric (HCl), nitric (HNO_3) and sulphuric (H_2SO_4). Depending on the concentration and raw material used, the effect produced may vary. Several authors have reported enhanced migration of nutrients (N or P) to process water for acid-promoted HTC processes [9].

- **Strong mineral bases**

Some of the strong bases are used with calcium oxide (CaO) and sodium hydroxide (NaOH); these compounds can influence the HTC in a way that increases the alkalinity of the medium, which favours the decomposition of organic matter and the reduction of pollutants improving the quality of HC since CaO increases the C content, promotes the formation of porous structures and reduces the formation of polycyclic aromatic hydrocarbons (PAHs), while NaOH improves the stability of HC and reduces SO_x and NO_x emissions. In addition, both catalyse key chemical reactions, increasing the Solid Yield (SY) of HC and facilitating CO_2 capture [10–12].

On the other hand, KOH has been demonstrated to inhibit the formation of secondary HCs; in relation to porosity of derived activated carbons, pyrolysis of HCs made using KOH can show improved microporosity development [13].

- **Metal chlorides**

Metal chlorides such as $FeCl_3$ are widely used due to their low cost; this is because the catalytic activity of these compounds is good due to the acid formations, decarboxylation and decarbonylation reactions are enhanced. They also reduce the pH of the medium, which is favourable because it aids the oxidation of the metal compounds present, although they can lead to higher ash content [14].

- **Sulphates, persulphates and sulphite salts**

These compounds affect the carbonization rate and structure of the carbon. Persulphates such as activated Fe(II) improve yields by promoting the decomposition of the initial raw material. Sulphites such as $Na_2S_2O_4$ improve char quality by preventing the formation of melanoidins [15].

- **Nitrate salts**

These compounds such as $Mg(NO_3)$ or Fe_2O_3 facilitate the thermal removal of biomasses such as sewage sludge and improve the decarboxylation process by obtaining a reduction of the organic nitrogen contained in the HC, as it is directed towards the liquid phase, improving its quality [16, 17].

- **Oxidation**

Oxidation, through agents such as hydrogen peroxide, decomposes the biopolymer structure, producing low molecular weight compounds [18]. This improves the dehydration efficiency and process competence, although it may reduce the hydrocarbon yield.

- **Other inorganic catalysts**

In addition to all the inorganic catalysts described above, there are other catalysts such as $NaHCO_3$ that improve the structure and porosity of HC, facilitating the decomposition of organic matter and reducing VOC, S and N [19]. Others, such as metal carbonates, are able to increase the pressure in the system, facilitating water income and improving biomass conversion. As for metallic salts, ammonium phosphate has been found to be the most effective in improving High Heating Value (HHV) and SY of HC [20]. In addition, combining different salts increases the ash content but reduces the H and O ratio. Other niobium-based catalysts [21] and zeolites [22] have also shown improvements in the energy content and quality of HC.

2.4 Questions

1. What is the main function of water in the HTC process?

   (a) To act as a source of oxygen as it is necessary for the HTC process
   (b) To function as a solvent and reagent
   (c) To generate high temperatures
   (d) Produce methane gas

2. At what temperature range is the organic matter or precursor residue subjected to in the HTC?

   (a) 80–120 °C
   (b) 180–260 °C
   (c) 300–400 °C
   (d) 50–150 °C

3. Which of the following compounds is not typically a product of the liquid phase (process water) in HTC?

   (a) Carboxylic acids
   (b) Alcohols
   (c) Hydrochars
   (d) Aldehydes

4. Which chemical reaction is represented by the formula: $C_6H_{12}O_6 \rightarrow 6C + 6H_2O$?

   (a) Hydration
   (b) Dehydration
   (c) Decarboxylation
   (d) Polymerization

5. Which of the following is a gas commonly produced during HTC?

   (a) Oxygen
   (b) Carbon dioxide
   (c) Nitrogen
   (d) Argon

6. Which stage of HTC involves the loss of hydroxyl groups?

   (a) Hydration
   (b) Dehydration
   (c) Polymerization
   (d) Aromatization

7. What is the role of intermediates in HTC?

   (a) They are the end products of the process

   (b) They act as secondary reagents in subsequent steps

   (c) They facilitate the dissolution of the gaseous products

   (d) They inhibit dehydration reactions

8. During aromatization in HTC, what types of compounds are mainly formed?

   (a) Linear compounds

   (b) Aromatic compounds

   (c) Aliphatic compounds

   (d) Organometallic compounds

9. What is meant by "secondary hydrochar" in the context of HTC?

   (a) A type of gas produced during carbonization

   (b) A solid generated due to the liquid composition of the medium

   (c) A gaseous by-product

   (d) A type of hydrochar with lower density

10. In the decomposition process of organic matter in HTC, which compounds are mainly removed during decarboxylation?

   (a) Hydrocarbons

   (b) Carbon dioxide and carbon monoxide

   (c) Oxygen and hydrogen

   (d) Water and alcohols

11. Which reaction involves the joining of organic monomers to form larger polymers?

   (a) Hydration

   (b) Condensation

   (c) Polymerization

   (d) Dehydration

12. What type of structure is formed during C–C bond formation reaction in HTC?

   (a) Two-dimensional structures

   (b) More stable three-dimensional structures

   (c) Single-crystal structures

   (d) Metallic structures

13. What is the main purpose of adding catalysts in HTC process?

   (a) To reduce the reaction temperature

   (b) To increase the stability of the hydrochar

   (c) To improve the efficiency and selectivity of the process

   (d) To generate gaseous products

14. Which of the following is not an important factor in the selection of a catalyst for HTC?

(a) Composition of the organic substrate
(b) Reaction conditions such as temperature and pressure
(c) Colour of the catalyst
(d) Residence time

15. What is the main effect of the pH of the medium on the activity of the catalysts during HTC?

(a) It determines the reaction temperature
(b) It influences the stability and selectivity of the catalyst
(c) It has no effect on the process
(d) It controls the gas production

16. Which of the following catalysts is a strong mineral acid used in HTC?

(a) Acetic acid
(b) Citric acid
(c) Sulphuric acid
(d) Lactic acid

17. Which of the following is not an inorganic catalyst used in HTC?

(a) Sodium hydroxide (NaOH)
(b) Iron chloride (FeCl$_3$)
(c) Acetic acid
(d) Sodium sulphate (Na$_2$SO$_4$)

18. What is the main function of alcohols as catalysts in the HTC process?

(a) To increase CO$_2$ production
(b) To facilitate dehydration and repolymerization
(c) They act as non-polar solvents
(d) Reduce hydrochar production

19. What effect do metal chlorides such as FeCl$_3$ have on the HTC process?

(a) They decrease the formation of gaseous products
(b) They increase the pH of the medium
(c) They reduce the ash content in the hydrochar
(d) They favour decarbonylation and dehydration reactions

20. Which compound is used as an organic catalyst in HTC to improve the micro- and mesoporous structure of hydrochar?

(a) Citric acid
(b) Sodium hydroxide
(c) Magnesium nitrate
(d) Formaldehyde

21. Which of the following statements is correct about the effects of strong mineral bases as catalysts in HTC?

(a) They increase the formation of volatile organic compounds

    (b) They reduce the yield of hydrochar

    (c) They improve the yield of hydrochar and reduce the emission of toxic compounds

    (d) Inhibit polymerization reactions

22. What role does oxidation play in the efficiency of dehydration in the HTC process?

    (a) It reduces the formation of low molecular weight compounds

    (b) It promotes the formation of low-quality hydrochar

    (c) Improves dehydration efficiency and process competence

    (d) It has no impact on dehydration

23. What effect does the addition of sulphates have on the HTC process?

    (a) It decreases the carbonization rate

    (b) It improves the structure and yield of hydrochar

    (c) It reduces CO_2 formation

    (d) It increases the pH of the medium

24. What is the effect of nitrates such as $Mg(NO_3)$ on the HTC process?

    (a) They increase methane formation

    (b) They improve the thermal removal of biomass and reduce the nitrogen content in the hydrochar

    (c) They decrease dewatering efficiency

    (d) They promote the formation of aromatic compounds

25. What is the characteristic of niobium and zeolite-based catalysts in HTC?

    (a) They are very expensive and are not used in industry

    (b) They do not affect the quality of hydrochar

    (c) They improve the final quality of the hydrochar

    (d) They reduce the reaction rate

26. Which catalyst is most effective in promoting cellulose decomposition in the HTC process?

    (a) Zinc chloride ($ZnCl_2$)

    (b) Citric acid

    (c) Methanol

    (d) Sodium sulphate (Na_2SO_4)

Competing Interests The authors have no conflicts of interest to declare that are relevant to the content of this chapter.

References

1. X. Lu, P.J. Pellechia, J.R. Flora, N.D. Berge, Influence of reaction time and temperature on product formation and characteristics associated with the hydrothermal carbonization of cellulose. Biores. Technol. **138**, 180–190 (2013). https://doi.org/10.1016/j.biortech.2013. 03.163

2. K.Q. Tran, T.T. Trinh, New insights into the hydrothermal carbonization process of sewage sludge: a reactive molecular dynamics study. Fuel **361**, 130692 (2024). https://doi.org/10.1016/ j.fuel.2023.130692

3. Q. Li, Y. Liu, Y. Wang, Y. Chen, X. Guo, Z. Wu, B. Zhong, Review of the application of biomass-derived porous carbon in lithium-sulfur batteries. Ionics **26**, 4765–4781 (2020). https://doi.org/ 10.1007/s11581-020-03694-3/Published

4. A. Funke, F. Ziegler, Hydrothermal carbonization of biomass: a summary and discussion of chemical mechanisms for process engineering. Biofuels Bioprod. Biorefin. **4**(2), 160–177 (2010). https://doi.org/10.1002/bbb.198

5. Y. Shen, J. Sun, Y. Yi, B. Wang, F. Xu, R. Sun, 5-Hydroxymethylfurfural and levulinic acid derived from monosaccharides dehydration promoted by InCl3 in aqueous medium. J. Mol. Catal. A Chem. **394**, 114–120 (2014). https://doi.org/10.1016/j.molcata.2014.07.007

6. O.S. Djandja, R.K. Liew, C. Liu, J. Liang, H. Yuan, W. He, Y. Feng, B.G. Lougou, P.G. Duan, X. Lu, S. Kang, Catalytic hydrothermal carbonization of wet organic solid waste: a review. Sci. Total Environ. **873**, 162119 (2023). https://doi.org/10.1016/j.scitotenv.2023.162119

7. J.G. Lynam, M.T. Reza, V.R. Vasquez, C.J. Coronella, Effect of salt addition on hydrothermal carbonization of lignocellulosic biomass. Fuel **99**, 271–273 (2012). https://doi.org/10.1016/j. fuel.2012.04.035

8. S. Kang, X. Li, J. Fan, J. Chang, Solid fuel production by hydrothermal carbonization of black liquor. Biores. Technol. **110**, 715–718 (2012). https://doi.org/10.1016/J.BIORTECH. 2012.01.093

9. A. Sarrion, E. Diaz, M.A. de la Rubia, A.F. Mohedano, Fate of nutrients during hydrothermal treatment of food waste. Biores. Technol. **342**, 125954 (2021). https://doi.org/10.1016/J.BIO RTECH.2021.125954

10. J.X. Wang, S.W. Chen, F.Y. Lai, S.Y. Liu, J.B. Xiong, C.F. Zhou, H.J. Huang, Microwave-assisted hydrothermal carbonization of pig feces for the production of hydrochar. J. Supercrit. Fluids **162**, 104858 (2020). https://doi.org/10.1016/J.SUPFLU.2020.104858

11. T. Liu, L. Tian, Z. Liu, J. He, H. Fu, Q. Huang, H. Xue, Z. Huang, Distribution and toxicity of polycyclic aromatic hydrocarbons during CaO-assisted hydrothermal carbonization of sewage sludge. Waste Manage. **120**, 616–625 (2021). https://doi.org/10.1016/J.WASMAN. 2020.10.025

12. C. He, K. Wang, A. Giannis, Y. Yang, J.Y. Wang, Products evolution during hydrothermal conversion of dewatered sewage sludge in sub-and near-critical water: effects of reaction conditions and calcium oxide additive. Int. J. Hydrogen Energy **40**(17), 5776–5787 (2015). https:// doi.org/10.1016/J.IJHYDENE.2015.03.006

13. V. Tkachenko, N. Marzban, S. Vogl, S. Filonenko, M. Antonietti, Chemical insights into the base-tuned hydrothermal treatment of side stream biomasses. Sustain. Energy Fuels **7**(3), 769–777 (2023). https://doi.org/10.1039/d2se01513g

14. X. Lu, X. Ma, Z. Qin, X. Chen, W. Yue, Co-hydrothermal carbonization of sewage sludge and swine manure: hydrochar properties and heavy metal chemical speciation. Fuel **330**, 125573 (2022). https://doi.org/10.1016/j.fuel.2022.125573

15. N. Yang, S. Yang, X. Zheng, Inhibition of Maillard reaction during alkaline thermal hydrolysis of sludge. Sci. Total Environ. **814**, 152497 (2022). https://doi.org/10.1016/J.SCITOTENV. 2021.152497

16. K. Mu, Q. Zhang, G. Luo, J. Han, L. Qin, B. Zhao, W. Chen, L. Yi, Role of iron conditioners on organics evolution in overall process of sludge hydrothermal carbonization followed by pyrolysis. Renew. Energy **198**, 169–175 (2022). https://doi.org/10.1016/J.RENENE.2022. 08.031

17. Y. Zhou, J. Remón, X. Pang, Z. Jiang, H. Liu, W. Ding, Hydrothermal conversion of biomass to fuels, chemicals and materials: a review holistically connecting product properties and marketable applications. Sci. Total Environ. **886**, 163920 (2023). https://doi.org/10.1016/J.SCITOTENV.2023.163920

18. M. He, X. Zhu, S. Dutta, S.K. Khanal, K.T. Lee, O. Masek, D.C. Tsang, Catalytic co-hydrothermal carbonization of food waste digestate and yard waste for energy application and nutrient recovery. Biores. Technol. **344**, 126395 (2022). https://doi.org/10.1016/J.BIORTECH.2021.126395

19. B. Du, Z. Yu, Y. Tian, X. Ma, Effects of baking soda on Co-hydrothermal carbonization of sewage sludge and Chlorella vulgaris: improved the environmental friendliness of hydrochar incineration process. J. Environ. Chem. Eng. **9**(6), 106404 (2021). https://doi.org/10.1016/J.JECE.2021.106404

20. S. Weihrich, X. Xing, Screening of synergetic catalytic effects of salts dominant in sewage sludge on corn stalk derived hydro and biochar. BioEnergy Res. **14**, 978–990 (2021). https://doi.org/10.1007/s12155-020-10192-x/Published

21. S. Kang, R. Miao, J. Guo, J. Fu, Sustainable production of fuels and chemicals from biomass over niobium based catalysts: a review. Catal. Today **374**, 61–76 (2021). https://doi.org/10.1016/J.CATTOD.2020.10.029

22. C. Peng, Y. Zhai, A. Hornung, C. Li, G. Zeng, Y. Zhu, Promoting effect of ZSM-5 catalyst on carbonization via hydrothermal conversion of sewage sludge. ACS Sustain. Chem. Eng. **6**(7), 9461–9469 (2018). https://doi.org/10.1021/acssuschemeng.8b02012

Chapter 3
Types of Wastes that Can Be Used in HTC

3.1 What is a Waste?

In this chapter, the main definitions and classifications according to different criteria will be presented. It is important to differentiate and to understand the main particularities or features of different wastes (according to their origin) that can be used in HTC, as their characteristics and composition will play a key role for the suitable selection of specific methods or techniques.

There are plenty of definitions for the term waste, like the following:

Every substance or object that an owner discards (voluntarily or as an obligation) or intends to discard [1].

This definition is broad and vague, as not every waste can be considered as such. Additionally, every substance or surplus that can be considered a waste might not necessarily considered as such from a legal point of view. In that sense, a general classification of wastes will be shown, to better understand this concept.

This classification might vary according to specific standards for each region. Thus, in the European Union, for instance, different guidelines for obligatory compliance are established for its member states. According to them, the concept of a waste is defined and, subsequently, its management.

Thus, according to the standard [1], once a waste is defined, it is necessary to establish what would not be considered a waste according to this norm:

- Flue gas released to the atmosphere.
- Land, including unexcavated contaminated soil and buildings in permanent contact with land.
- Uncontaminated soil, as well as other natural material that is excavated during building activities (as long as it is certain that this material will be used for building purposes at the same site where it was extracted).
- Radioactive waste.
- Decommissioned explosives.

B. Ledesma Cano et al., *Introduction to Hydrocarbonization*,
SpringerBriefs in Applied Sciences and Technology,
https://doi.org/10.1007/978-3-031-70039-2_3

- Faecal matter (if they are not included in specific regulations), straw and agricultural or winegrowing material that is used for typical purposes like agriculture, forestry or energy production from this biomass through methods or technologies that are not harmful to the environment or the human body.

Other elements are excluded as wastes due to the fact that they present their own specific regulations, like the following:

- Wastewater.
- Animal by-products, including processed products covered by 1774/2002 Regulation (CE), except for those intended for incineration, landfill or those used in gas or composting plants.
- Animal carcasses from animals that have not been slaughtered, including practices to eradicate epizootic, which are managed according to the 1774/2002 Regulation (CE).
- Wastes resulting from prospecting, extraction, management or storage of mineral resources, as well as the working of quarries covered by the 2006/21/CE European Directive (15 March 2006), about waste management of extractive industries (according to DO L 102 de 11.4.2006, p. 15).

Also, it is necessary to distinguish several aspects. For instance: what is the difference between a by-product and a waste? What can be considered as a hazardous waste? The abovementioned regulation also includes some considerations about it, with the following definitions:

Definition. Hazardous waste: A waste that presents one or several characteristics included in this regulation ([1] Annex III), remarking some of the following:

- Oxidizing: It applies to substances and compounds that present highly exothermic reaction in contact with other substances, particularly flammable ones.
- Explosive: Those substances and compounds that can explode in contact with a flameo or those that are more sensitive to impacts and friction than dinitrobenzene.
- Easily flammable: It applies to substances and liquid formulae with a flash point below 21 °C (including extremely flammable liquids); substances that can be heated and burn in contact with air at room temperature; solid substances that can burn with a brief contact with an ignition source, continuously burning once this source is removed; gas compounds that are flammable in air at normal pressure; substances and compounds that, in contact with water or steam, generate considerable amounts of easily flammable gas.
- Flammable: For substances or a liquid formula that present a flash point above or equal to 21 °C and lower than or equal to 55 °C.
- Irritant: It applies to non-corrosive substances that can provoke inflammatory reactions due to immediate, constant or repeated contact with the skin or mucous membrane.
- Toxic: It applies to substances or compounds (including very toxic ones) that, due to inhalation, ingestion or skin penetration, can provoke serious risks, acute or chronic, or even death.

- Corrosive: It applies to substances or compounds that can destroy living tissues in contact with them.
- Ecotoxic: For those wastes that present or might present immediate or deferred risks for the environment.

Definition. By-products: A substance or object, resulting from the production process, whose primary aim is not its production, could be considered as a by-product and not as a waste according to the abovementioned definitions and requirements, as long as these products comply with the following conditions:

- Its subsequent use is guaranteed.
- They can be directly used without any further process that is different from typical industrial practices.
- They are produced as an integral part of the production process.
- Its final use is legal, that is, the by-product complies with all the applicable requirements concerning the environment and public health, not provoking adverse effects on the environment or human health in general.

Therefore, according to the previous definition, not every product or substance considered to be a by-product should be considered as a waste and, consequently, would be potentially excluded from its possible energy conversion through techniques such as HTC or co-HTC.

As previously indicated, this classification and definition of waste are given according to European standards. Nevertheless, considering a possible classification to get a general idea, a simpler and more general classification can be obtained to present a first approach about this concept:

According to its nature: Hazardous and non-hazardous wastes. The former are those that comply with the abovementioned conditions. Therefore, non-hazardous wastes would be those that do not comply with the previous concept.

According to its source: Wastes can be classified as domestic, commercial or industrial wastes. Moreover, there are wastes that, due to their special characteristics, have to comply with a specific regulation or treatment.

- Domestic wastes: They are wastes that are generated in households as a consequence of domestic activities, as well as those produced in industries and service sector. Equally, wastes generated by electrical or electronic devices, clothes, batteries, accumulators or furniture (as well as wastes and debris due to minor Works) are also included. Moreover, those wastes produced during street cleaning, in green areas or beaches, dead pet animals or abandoned vehicles, among others, are considered.
- Commercial wastes: These wastes are generated on account of the typical commercial activity (wholesale or retail) of restaurant or bar services, offices or markets, as well as the remaining activities related to the service sector.
- Industrial wastes: They are both hazardous and non-hazardous wastes, produced in different industrial activities such as manufacturing, transformation, consumption, cleaning or maintenance processes, among others, excluding emissions to the atmospheres.

A fourth level can be given, included in the abovementioned regulation, that would embrace several aspects of the previous classification:

- Biowaste: This is a biodegradable waste with different origins: from gardens or parks; food waste from households, restaurants or retail establishments; or wastes from food processing plants.

Considering the above, it would be possible to identify (at least approximately) at different levels. However, the aim of this chapter is to establish the advantages of HTC and co-HTC processes and the characteristics of the products obtained. Thus, this technique was mainly focused on the valorization and transformation of biomass waste. Therefore, it is necessary to define the concept of biomass and its different kinds.

3.1.1 Organic Waste Sources

As previously explained, HTC is mainly used for biomass, even though not only biomass products are appropriate for this process. Thus, a definition of biomass, as well as the example of several suitable materials is necessary to understand HTC and co-HTC processes.

Biomass can be defined as every material of biological origin which has not undergone mineralization processes, like petroleum, gas or coal. Specifically, biomass is defined as every biological material where photosynthesis processes have directly or indirectly taken place, being generated recently compared to mineralization processes. According to this definition, biomass can be classified as observed in Fig. 3.1.

- Natural biomass: It can be defined as the biomass directly obtained from nature, except those intended for human consumption. For example, forestry waste is obtained in the clearing of forests.
- Residual biomass: It is obtained in industrial, agricultural and forestry processes, for instance during fruit tree pruning, nutshell removal in industries or harvesting (where agricultural wastes such as corncob can be obtained). In this category, other wastes can be included, such as municipal solid waste or solid waste from livestock holdings, slurry, manure or sewage sludge obtained from wastewater purification plants.
- Energy crops: These are crops used for fuel production, such as biodiesel or biomethane.

Therefore, as described in Chap. 1 about the possible research lines related to the products obtained in HTC, different biomasses are used as starting raw materials. Among them, lignocellulosic biomass plays an important role, as the majority of biomass is agricultural or forestry wastes, which mainly consist of these compounds.

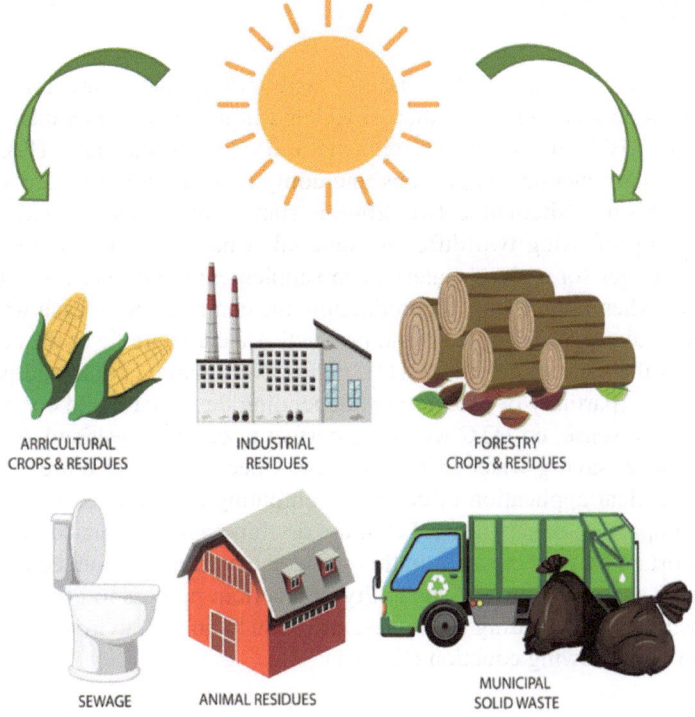

ARRICULTURAL
CROPS & RESIDUES

INDUSTRIAL
RESIDUES

FORESTRY
CROPS & RESIDUES

SEWAGE   ANIMAL RESIDUES

MUNICIPAL
SOLID WASTE

Fig. 3.1 Origin and biomass types

Also, as explained in Chap. 1, the final products obtained through HTC are mainly liquid and solid products (HC). Their characteristics will vary depending on multiple factors, such as processing or, especially, the starting material used for this purpose. Thus, according to the previous section, several materials are potentially suitable for HTC process. However, not all wastes have been assessed to be used in HTC, mainly focused on their viability to obtain energy products, their characterization of some properties such as toxicity or their economic feasibility.

In any case, as water plays an essential role in HTC, it seems interesting the use of wet biomass waste for this kind of process. In that sense, wastes with a great potential in this context could be sewage sludge, some food waste or agricultural waste without any previous drying process and high moisture levels. In these cases, the use of water is usually reduced, only required to adjust the optimum organic material/water ratio to obtain higher yields for specific purposes, as explained in further chapters.

Nevertheless, the abovementioned reasoning does not mean that dry biomass is not suitable for HTC. It is true that water addition would be a challenge, especially for processes at industrial scale, where this could be an expensive and unsustainable step compared to the use of wet biomass. Nonetheless, the use of an intermediate alternative has aroused interest in the scientific community: the combined use of dry and wet biomass to obtain a particular case of HTC, specifically called co-HTC.

3.2 Co-HTC Process

Accordingly, the influence of the kind of waste on the final characteristics of the main products obtained in HTC (that is, HC and liquid) is considerable. Recently, a new trend in HTC has emerged, with the aim of using the same HTC process (including similar chemical conditions, including temperature, pressure and time). However, in this new alternative, two different starting materials are used. The main reason of using of using two different materials (mainly wastes) is the fact that diverse advantages for a given material can supplement the properties of the other. For instance, when it comes to HC production, the contribution of each waste could balance the final properties of this solid product. A clear example of the advantages of co-HTC is the water required for HTC. Thus, wet biomass could add part of the water, reducing (partially or even totally, depending on the case and conditions) its addition. In that sense, co-HTC would be perfect to combine wet and dry biomass for this purpose, saving water in this process. Specifically, there are studies that prove the practical application of co-HTC combining dry biomass with grass [2]. Hence, a comparison of separate HTC of the dry biomass was carried out, pointing out the important savings in water. In this sense, the concept of atom economy (AE) is interesting, as it relates to the capability of a certain process to use all the atoms included as reagents, avoiding the release of products in the environment. Thus, as observed in the following equation (Eq. 3.1):

$$\text{Atom economy } (\%) = \frac{\sum \text{MW}_{\text{desired products}}}{\sum \text{MW}_{\text{reactants}}} \times 100 \qquad (3.1)$$

This parameter can be expressed in % (or as a ratio between 0 and 1). The atom economy of a process would be higher (up to 100%, the optimum value where the reactants are completely used for the generation of desired products) if the products are desired, that is, used for energy purposes, as added-value materials, etc. The higher AE is, the lower the direct environmental impact is, as no by-products or wastes are generated in this process and, subsequently, evolved to the environment. Nevertheless, the life cycle assessment of a certain process should be taken into account, where the energy costs (among other factors) could imply indirect environmental impacts. In any case, the use of combined biomass in co-HTC could be an interesting process with a high atom economy compared to the direct combustion/pyrolysis of the corresponding wastes.

Compared to HTC, co-HTC is an innovative technology, with references from 2016. Nevertheless, this kind of HTC has gained interest, as observed in Fig. 3.2, where an increasing trend in the number of publications related to this subject has been more noticeable compared to traditional HTC.

As observed in this figure, the increase in publications about co-HTC has been considerable since 2017, reaching similar results in published articles in the last years for both HTC and co-HTC. It is due to the increasing interest in this technique due to its versatility. Thus, in recent years, the scientific community has resorted to

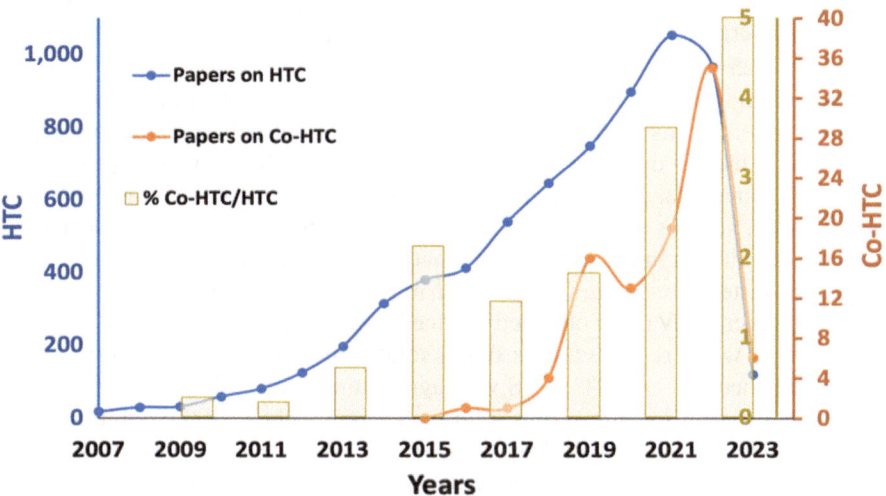

Fig. 3.2 Comparison of articles about HTC and co-HTC in 2016–2023

co-HTC as a solution for the suitable management of some wastes whose removal or transformation would be highly complex otherwise. According to the kind of raw material selected to carry out the combined HTC process, which should be beneficial, the researchers have focused on the following options:

- **Co-HTC biomass-biomass**: This trend represents the highest percentages of studies focused on co-HTC, with 56% of total published articles. In that sense, the valorization of two biomass wastes is the aim research line.
- **Co-HTC biomass-coal**: This option is interesting, with 17.6% of total publications. Thus, these studies are focused on coal wastes combined with biomass.
- **Co-HTC biomass-PVC**: In this case, this alternative is focused on the dechlorination of PVC through hydrocarbonization with lignocellulosic biomass. Articles about these subjects represent 18.7% in total.
- **CO-HTC coal–PVC**: Finally, 3% of these studies deal with plastic wastes combined with fossil fuel wastes in HTC.
- **Reviews**: Given the increasing interest in this combined technique, four review works have already been published (representing 4.4% of total articles about co-HTC).

For instance, an example of and specific application of co-HTC could be the production of a marketable and sustainable solid biofuel with improved characteristics. Moreover, scientists continue with the search for other additional advantages related to co-HTC. Thus, 13% of total articles related to co-HTC are focused on chlorine removal from plastics. As a synthetic plastic polymer, PVC has a high energy content (15–40 MJ/kg) that can be turned into utilizable fuels or power through a conversion system of energy wastes to energy [3]. In this sense, there are several publications given the relevance of such a worrying topic, as worldwide PVC production exceeds 40 million tonnes per year, expecting an increase to above 60 million tonnes by 2025 [4].

Another interesting alternative for this purpose is the use of sewage sludge (SS). The production of this waste in wastewater treatment plants has continuously been increasing in the last decades, whereas its disposal in landfills has been proven as a serious environmental problem, looking for sustainable choices for the management of this waste, such as pyrolysis or HTC, among others [5, 6]. Co-HTC technique has attracted interest in this sense, given that it does not require the previous drying process of wastes like sewage sludge, which makes HTC and co-HTC suitable for this waste. However, HHV of the obtained biochar is still relatively low due to the high ash content in SS, which restricts the use of this solid waste. Nevertheless, several studies have established that co-HTC of SS with agricultural waste could provide HC with high-quality fuels [7], compared to solid fuels directly obtained through traditional HTC. Thus, hydrochar derived from co-HTC had a higher HHV (increased by 2.6 times). As observed in Fig. 3.3, where the van Krevelen diagram is shown (where different solid products can be found according to their H/C and O/C ratio), HC obtained from co-HTC is principally located in the lignite and coal region (expecting a similar behaviour), which corroborates the higher HHV compared to typical HTC, improving the heating value of biomass [8]. More importantly, a synergistic effect of combined co-HTC of this lignocellulosic biomass and sewage sludge was observed during this process. Thus, concentration and bioavailability of heavy metals in HC decreased through complexation, precipitation, adsorption or other procedures [9]. Another advantage of the co-HTC in sewage sludge is the possible reduction in heavy metal content in the resulting hydrochar, due to the relatively high content of Cd (up to 5 mg/kg), Cr (with a wide range depending on the sample, from around 20 to 3000 mg/kg), Cu (up to 600 mg/kg), Zn (from around 1 to above 1000 mg/kg), Ni (up to 271 mg/kg) or Pb (exceeding 500 mg/kg in some cases), among others, in SS. In that sense, the use of the resulting HC from a direct HTC for specific purposes would be inadequate, as in the case of its possible use as a soil conditioner (as these heavy metals can transfer to crops). Also, possible high levels of ash would make its use as a biofuel difficult. For this purpose, it has been proven the effectiveness of its co-HTC with biomass, proving a considerable reduction in Cu, Zn, Cr and Ni [10].

3.2.1   Synergistic Effect

As evidence of the potential benefits of co-HTC, several trends in recent research works have been dealt. As previously explained, the advantages of the combined use of wastes in co-HTC, as in the case of SS and biomass waste, have been previously commented. Also, in order to assess the synergistic (and positive) effect of this combination, the synergistic coefficient or index can be calculated. It can be defined

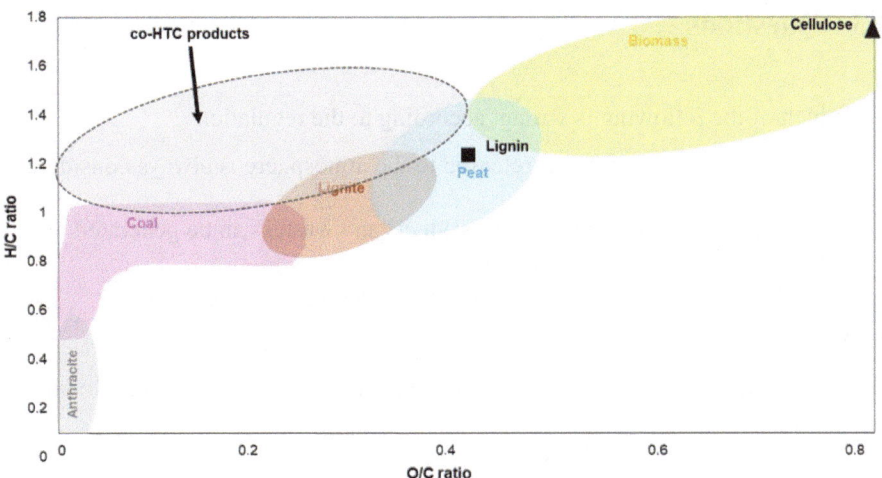

Fig. 3.3 Van Krevelen diagram, including several solid fuels and the main region for HC obtained from co-HTC

as a way to quantify the effect produced during co-HTC, assessing the improvement compared to the traditional HTC use of one single waste (expressed in %).

The origin of the research about the synergistic effect was carried out recently [11]. In this study, 33 works were reviewed to discuss the synergistic index. However, it was not calculated in any case. Additionally, different articles where the authors offered valid data were included, with the possible comparison of different parameters of HC depending on different variables (such as blending ratio of wastes, temperature, etc.). Thus, the synergistic coefficient can be defined according to the following equation (Eq. 3.2) [12]:

$$\text{synergistic coefficient} = \frac{\text{Experimental value} - \text{Calculated value}}{\text{Calculated value}} \times 100 \quad (3.2)$$

The previous equation is used by most researchers, where the theoretical value is obtained by the parameters of a certain raw material and its ratio in the mixture. The introduction of this coefficient aims to describe the different between experimental values and those obtained theoretically, in order to determine the possible synergistic effect. This coefficient is related to the quality and quantity of hydrochar obtained from co-HTC. According to the literature, this index represents the intensity of the synergistic effects given during a certain HTC reaction [12]. For the corresponding synergistic effect, some representative parameters related to hydrochars are usually studied, such as the yield or HHV, for instance. As a result, different effects can be found according to the previous equation: synergistic (positive coefficient), antagonistic (negative coefficient) or additive (null coefficient) [7].

3.3   Questions

1. Which of the following is wrong, according to the regulations?

   (a) In a certain process, gas released to the atmosphere is always considered a waste
   (b) After a process, products, by-products and wastes can be generated
   (c) Some wastes can be hazardous
   (d) There are different types of wastes, depending on many factors

2. Why is sewage sludge a suitable waste for HTC?

   (a) Its moisture level is considerable, which could reduce water addition in this process
   (b) Due to its high ash content
   (c) On account of its low density
   (d) All of the above

3. Which of the following is correct about co-HTC?

   (a) It is the use of carbon monoxide in HTC
   (b) Its use can improve some properties in the resulting hydrochar or liquid
   (c) The use of two wastes in HTC normally presents an antagonistic effect
   (d) a and b are correct

4. Atom economy of a process usually ranges from:

   (a) 0 to 100%
   (b) 0 to 1
   (c) 0 to infinity
   (d) a and b are correct

5. What is the most common combination of raw materials in co-HTC, according to the literature?

   (a) Biomass-biomass
   (b) Biomass-PVC
   (c) Biomass-coal
   (d) None of the above

6. Regarding the Van Krevelen diagram, which of the following is true?

   (a) It is an interesting diagram where a qualitative comparison of HHV can be accomplished
   (b) It represents H/C versus N/C ratios of solid fuels
   (c) Different regions of typical solid fuels can be found
   (d) a and c are correct

7. If the synergistic coefficient is positive, it means:

(a) There is an antagonistic effect
(b) There is an additive effect
(c) There is a synergistic effect
(d) None of the above

Competing Interests The authors have no conflicts of interest to declare that are relevant to the content of this chapter.

References

1. P. European, DIRECTIVE 2008/98/EC of the European Parliament and of the Council of 19 November 2008 on waste and repealing certain Directives (2008). Available online: https:// eur-lex.europa.eu/legal-content/EN/TXT/?uri=celex%3A32008L0098
2. R. García-Morato, S. Román, B. Ledesma, C. Coronella, Co-hydrothermal carbonization of grass and Olive stone as a means to lower water input to HTC. Resources **12**(7), 85 (2023)
3. Y. Qi, J. He, F.R. Xiu, W. Nie, M. Chen, Partial oxidation treatment of waste polyvinyl chloride in critical water: preparation of benzaldehyde/acetophenone and dechlorination. J. Clean. Prod. **196**, 331–339 (2018). https://doi.org/10.1016/j.jclepro.2018.06.074
4. E.K. Petrović, L.K. Hamer, Improving the healthiness of sustainable construction: example of polyvinyl chloride (PVC). Buildings **8**(2), 28 (2018)
5. Q. Dai, Q. Liu, X. Zhang, L. Cao, B. Hu, J. Shao, F. Ding, X. Guo, B. Gao, Synergetic effect of co-pyrolysis of sewage sludge and lignin on biochar production and adsorption of methylene blue. Fuel, **324**, 124587 (2022)
6. J.F. González, C.M. Álvez-Medina, S. Nogales-Delgado, Biogas steam reforming in wastewater treatment plants: opportunities and challenges. in *Energies*, vol. 16, issue 17 (Multidisciplinary Digital Publishing Institute (MDPI), 2023). https://doi.org/10.3390/en16176343
7. C. He, Z. Zhang, C. Ge, W. Liu, Y. Tang, X. Zhuang, R. Qiu, Synergistic effect of hydrothermal co-carbonization of sewage sludge with fruit and agricultural wastes on hydrochar fuel quality and combustion behavior. Waste Manage. **100**, 171–181 (2019)
8. M. Bardhan, T.M. Novera, M. Tabassum, M.A. Islam, M.A. Islam, B.H. Hameed, Co-hydrothermal carbonization of different feedstocks to hydrochar as potential energy for the future world: a review. J. Cleaner Prod. **298** (2021). https://doi.org/10.1016/j.jclepro.2021. 126734
9. Q. Lang, Y. Guo, Q. Zheng, Z. Liu, C. Gai, Co-hydrothermal carbonization of lignocellulosic biomass and swine manure: hydrochar properties and heavy metaltransformation behavior. Biores. Technol. **266**, 242–248 (2018)
10. A. Leghari, Y. Xiao, L. Ding, A. Raheem, A. Ryzhkov, G. Yu, Research advancements in nutrients and heavy metals, its speciation and behavior during hydro thermal carbonization of sludge—a critical review. Fuel **352** (2023). https://doi.org/10.1016/j.fuel.2023.129082
11. M. Bardhan, T.M. Novera, M. Tabassum, M.A. Islam, M.A. Islam, B.H. Hameed, Co-hydrothermal carbonization of different feedstocks to hydrochar as potential energy for the future world: a review. J. Clean. Prod. **298**, 12673 (2021)
12. Y. Lin, Y. Ge, H. Xiao, Q. He, W. Wang, B. Chen, Investigation of hydro thermal co-carbonization of waste textile with waste wood, waste paper and wast food from typical municipal solid wastes. Energy **210**, 118606 (2020)

Chapter 4
Practical Considerations

4.1 Parameters, Laboratory Techniques and Experiments Related to Hydrocarbonization

4.1.1 Experimental Procedure Carried Out to Perform HTC

Adequacy of biomass or starting precursor

Before designing the HTC process, biomass must first be prepared. Depending on the characteristics of the starting material and the final objective of the process, homogenization of the starting material or other properties for its use will be sought. For this purpose, one or several treatments will be chosen at the same time. Some of the steps that are usually followed in the laboratory are shown below.

1. Separation: If necessary, unwanted components should be separated before pre-treatment if there is high heterogeneity in the starting mass. For example, if olive pruning remains are selected, and only the small branches are desired, we must remove or separate the large branches at a glance.
2. Crushing: Once they have been separated the parts of the biomass either because they are unwanted or because they have different properties, such biomass is crushed to reduce the particle size. For this process, equipment such as mills, crushers or grinders can be used in order to improve the homogeneity and in turn the efficiency of the HTC.
3. Sieving: This process is used to obtain a more homogeneous sample. For this technique, sieves of different sizes are used to separate the biomass into fractions, giving uniformity to the sample.
4. Drying: This process is carried out when it is interesting to reduce the initial moisture content present in the biomass, to avoid possible complications in the process, in addition to facilitating its handling. Sometimes, if the biomass contains a high moisture content, the drying process is carried out before shredding, making it

easier to treat the biomass. In some applications, it may be preferable or necessary to use wet biomass, so this step would not be necessary. This would reduce costs and seek to maximize the efficiency of the process.

HTC's experimental procedure

Once the biomass is homogenized and uniform, the load to be introduced into the reactor is prepared, making the desired mixture of biomass and water according to the experimental conditions and the biomass to be used, and the weight of both the biomass and the water used is recorded.

The mixture is then introduced into the reactor and the reactor is hermetically sealed. The reactor is then placed in an oven at the desired temperature or subjected to a temperature-regulated heating system. This is where the HTC process takes place, the biomass is heated under self-pressure conditions for a period and particle decomposition occurs to generate the final products in both phases (liquid and solid), the gas phase is usually wasted because it is usually less than 5% of the total HTC.

When the reaction is complete, the reactor is cooled down. This can be done at room temperature by simply removing the reactor from the furnace or removing it from the heating system. It can also be done by a more abrupt cooling using a forced ventilation system or by placing it in ice water. For the separation of HC and PW, it is useful to use the vacuum filtration or suction filtration procedure (solid–liquid separation technique). The mixture is placed in a Buchner funnel (filtration funnel) with the filter paper attached to the bottom so that the holes in the funnel do not allow the HC to pass through while the PW is being filtered. In turn, a suction flask is needed in which the Buchner funnel is placed by means of an adapter. The vacuum is achieved by sucking the mixture with a pump to create a reduced pressure, thus speeding up the process.

Laboratory materials

The following is a description of the laboratory equipment required to perform the HTC technique at a laboratory scale.

- Balance: This is the instrument used in the laboratory to measure the amount of the precursors as well as the products obtained in the HTC process.
- Pressure reactor: Pressure reactors are used to carry out the HTC reaction itself, where the mixture of biomass and water is housed. The reactors must be highly reliable and safe, with the protection against corrosion. There are several types depending on how the heat is applied, whether they feature a stirring system, the size (from 0.25 ml to 100 L), the maximum operating temperatures and pressures (up to 500 °C and 50 bar- 345 bar, respectively), and the material of manufacture.
- Heating furnace: Used to provide heat to reactors that do not have a self-heating source. Also, it is used to dry the resulting HC for subsequent analysis.
- Vacuum filtration equipment: It consists of Buchner funnel, filtration flask (suction flask), Buchner funnel adapter, vacuum pump and filter paper.
- pH metre: It is used to obtain the pH and conductivity of the process water as it gives us relevant information to know how the medium is acidified.

Analysis

To evaluate the effectiveness of hydrocarbonization, it is crucial to carry out a series of analyses that shed light on the results of the process and the evolution of the products obtained. In this regard, characterization methods are used to understand in detail the changes that have occurred. These methods are described below together with concrete examples to facilitate their understanding.

4.1.2 Characterization of HC

The characterization techniques that are most commonly used for the analysis of hydrochars are presented below.

Higher heating value (HHV) analysis

The higher heating value (HHV) analysis is a fundamental measurement that determines the amount of heat released during the complete combustion of a fuel per unit mass.

Depending on how the calorific value of a solid, in this case HC, or of fuel is determined, a distinction can be made between higher and lower calorific values, depending on whether the determination is carried out in a closed or open vessel, respectively. Thus, when determining the calorific value of a fuel in a calorimeter, it can be observed that water is deposited on the walls of the calorimeter, which comes from the condensation of the water vapour produced during combustion or from the smoke contained in the fuel itself. This vapour gives up, on condensing, approximately 600 kcal (2508 kJ/kg) of water vapour, heat which must be added to the heat produced during combustion. The sum of both gives the gross calorific value (HHV) of the fuel.

On the other hand, if the same fuel is burned in an industrial furnace (system open to the outside), the water vapour in the fumes or combustion products does not condense, because the temperature at it which it is discharged is normally higher than 100 ºC (to avoid reaching the sulphuric dew point in fuels with a high sulphur content). In this case, the amount of energy released in the combustion of the fuel will be less than that obtained in a colorimeter, which is called the lower heating value (LHV) of the fuel. It differs from HHV, only and exclusively, in the heat of condensation of the water produced in the combustion or contained in the fuel.

Elemental analysis

The elemental analysis of carbon, hydrogen, oxygen, sulphur and nitrogen is carried out experimentally in an elemental analyser in which the N, C and H content of the HC can be determined.

The operation of this analyser is based on the determination, by thermal conductivity, of the concentration of the gases obtained in the combustion of the sample.

The analyser has several detectors, between the first and second detectors water is retained in a bed of magnesium perchlorate and between the third and fourth detectors CO_2 is retained in a tube; finally, N is determined by considering the helium content of the carrier gas by comparing its conductivity with that of a pure helium stream.

It is crucial to follow the standards UNE-EN 15104 to determine the content of C, H and N, and UNE-EN 15289 for the content of S, expressing the results in percentage by mass on a dry basis. Prior to analysis, each sample must be completely dry to remove hydrogen and oxygen from moisture, ensuring accurate results on a dry basis.

The percentage of C, H, N and S is calculated as an average of the values obtained in the tests carried out, these data being automatically calculated by the elemental analyser. On the other hand, the percentage of oxygen in the sample is obtained by difference, using the following equation (Eq. 4.1):

$$O = 100 - (C + H + N + S). \qquad (4.1)$$

From this analysis, carbon densification can be calculated which is associated with the increase in the proportion of carbon (%) found after the HTC process, which is seen when comparing the C (%) of the feedstock and that of the HC.

On the other hand, carbon capture (CC) (Eq. 4.2) represents the proportion in which the amount of C in the feedstock remains in the solid product and therefore considers the solid yield [1]:

$$cc\ (\%) = \frac{C_{\text{Hidrochar}}\ (\%)}{C_{\text{Feedstock}}\ (\%)} \times 100. \qquad (4.2)$$

H contents are indicative of water formation during combustion, which can reduce energy efficiency and cause condensation problems in combustion systems.

N in HC can produce nitrogen oxides (NO_x), especially if HC is to be used as fuel. These compounds are known to be air pollutants and contribute to acid rain and photochemical smog. In applications such as soil amendments, a lower N content may be preferable to avoid excess nutrients that could lead to soil and water pollution.

On the other hand, the S present in HC can generate sulphur dioxide (SO_2) during combustion, which is a toxic gas and a precursor of acid rain. As with N, if used in soil amendment, a low or zero proportion is desirable to avoid soil acidification and possible damage to crops.

The presence of O indicates a lower energy density of the HC. Lower oxygen means that the material is richer in C and therefore has a higher HHV. In addition, in applications such as adsorbent materials, a low O content can improve the adsorption capacity of certain organic pollutants.

Numerous studies have found that a low H/C ratio is indicative of high aromaticity of HC, while a low O/C ratio reveals fewer O-inclusive functional groups, and in turn both low ratios suggest improved HC stability as well as C sequestration potential. On the other hand, low $\frac{N+O}{C}$ content suggests low polarity [2] and can promote the reduction of water vapour and fumes during HC combustion if it were to be carried out [3]. Other studies have shown that the loss of oxygenated functional groups on the surface due to the decrease in H/C and O/C ratios increases the hydrophobicity of HC [4] which also makes it attractive for combustion since it absorbs less water from the environment.

Immediate analysis

The immediate analysis of a fuel comprises the determination of the moisture, ash, volatile matter and fixed carbon content of the fuel.

- Moisture

The amount of water, expressed as a percentage by weight, contained in a fuel. This characteristic is of particular importance in HC.

The amount of water retained by a biomass or final HC can vary and in fact varies within very wide limits. The amount of water contained in a solid.

The amount of retained water by a biomass or final HC can vary and in fact varies within very wide limits. The amount of water contained in a solid can be considered as consisting of two different types. The first is the so-called free, surface or added moisture and corresponds to the water mechanically bound to the fuel, during washing processes or by exposure to atmospheric agents. This moisture can be removed by depositing the fuel in thin layers inside a dry and aerated room.

It is obvious that some of the water contained in the fuel will remain in equilibrium with the ambient humidity and is called inherent; to remove it, it is necessary to heat the duel up to 110 °C, thus vaporizing the water.

- Ashes

The precursor biomass or the final product contains a certain amount of inert material matter (silica, aluminium, etc.) which after combustion gives rise to ash or slag. Ash is the solid residue, expressed as a percentage by weight, resulting from the complete combustion of a fuel.

Analogous to moisture, ash can be classified into two distinct types, inherent ash and adventitious ash.

Inherent ashes come from the non-combustible solid matter contained in the plants from which the HCs originate, while adventitious ashes have been added during the extraction processes.

They are determined by complete combustion of a pre-weighted fuel sample in a muffle furnace at a temperature of about 800 °C.

If the HC is intended for solid fuel, this characteristic is very important, together with the melting temperature of the ash, as both determine in part the value of the fuel for industrial uses. The amount of ash is a crucial sign of the possibility of slagging of the solid fuel. One HC with low ash content would burn cleaner and more efficiently, as the presence of inorganic minerals such as Si, K, Na, S, Cl, P, Ca, Mg and Fe can cause toxic emissions as well as fouling, slagging and corrosion in the combustion chambers.

- Volatile matter

Volatile matter is defined as the percentage loss in weight of a fuel (free of moisture) when heated in an inert atmosphere (absence of oxygen or other oxidant) for seven minutes at 900 °C.

Knowledge of this characteristic is, in some cases, indispensable to know the industrial applications for which a particular fuel is intended to be used.

It should be noted that when a solid fuel, which has been stripped of its moisture, is approached by a flame, the first thing that happens is the release of volatile matter and then the combustion of the volatile matter, which initiates the actual combustion process of the fuel.

- Fixed carbon

Fixed carbon (FC) (Eq. 4.3) means the actual combustible matter, expressed as a percentage, contained in a fuel.

$$FC = [100 - (\text{Moisture} + \text{Ash} + \text{Volatile})]. \quad\quad (4.3)$$

In solid fuels, or in HC in this case, once the volatile matter has been removed, what results is a solid residue called coke, which is made up of ash and the combustible matter that accompanies the fuel. The weight loss, expressed as a percentage, that occurs in the combustion coke is called fixed carbon; the residue that remains is the ash.

Each country follows its own standard setting out methods for the determination of total moisture, volatile matter and ash; however, the rationale for these methods is simple and very similar. Moisture is determined by difference weighing after drying the sample in an oven at 110 °C; volatile matter is determined by combustion of the sample in contact with air in an oven that reaches a temperature of 815 °C. Finally, the fixed carbon is determined by difference of above with the initial mass of the sample, expressing all values as a percentage. These methods, although simple, are tedious, especially when the number of samples is large.

From the above analysis, the fuel ratio (FR) can be calculated which is defined by the ratio of fixed carbon to volatile matter and is indicative of the stability of HC as a fuel. If the value obtained is less than or equal to 0.33, it indicates that HC is unstable and has a half-life of less than 100 years [5]. A high FR can increase the combustion temperature, thus improving flame stability and keeping the flame less aggressive, as well as reducing heat loss [6]. Conversely, a low FR reveals a high volatile matter content, which can generate unwanted soot in equipment [7].

Determination of biological contents

Determination of cellulose, hemicellulose and lignin content (that is, lignocellulosic content) can be determined by chemical methods although nowadays there are methods such as thermogravimetric analysis which are quicker and cheaper for such determination [8].

Scanning electron microscopy

The analysis of morphology of solid samples is carried out using a microscope that can observe and characterize organic and inorganic materials and provide results on the morphology (roughness, porosity and texture) of the HC to be studied. Therefore, scanning electron microscopy (SEM) is a very appropriate technique that consists of the incidence of an accelerated electron beam on a thick sample of material that is opaque to electrons. The beam is focused on the sample Surface in such a way that it sweeps the sample surface are used to vary the signal intensity in a cathode ray tube that moves in synchrony with the probe. In this way, there is a direct relationship between the position of the electron beam and the fluorescence produced in the cathode ray tube. The result is highly magnified topographic image of the sample.

If it is desired to extend the study of the elemental composition of the sample, the technique can be completed with qualitative and quantitative microanalysis by energy dispersive X-ray (EDX), which performs qualitative and quantitative analysis.

As an example, we will show some SEM studies that have been carried out in the laboratory, explaining the purpose of this technique.

In the following SEM images (see Fig. 4.1) made to different samples with different objectives. Figure (a) shows the nutshell after having performed a 3-h HTC, the purpose of this SEM is to see how the nutshell behaves superficially, which in this case is very porous surface. Figures (b) and (c) show SEM images of an activated carbon in the form of pellets at different magnifications after having performed a joint HTC with walnut shell for a duration of 8 h; the objective of this was to see how it behaves superficially observing small spherical formations of combination of compounds that are deposited on the sample.

From this technique, it must be emphasized that the surface composition will depend on which area of the sample is taken for photography, since superficially the sample is usually not uniform.

In turn, there are complementary techniques to SEM that offer a wide range of options to study the surface morphology of samples with different approaches and

a) nutshell after 3-hour HTC (500 x magnification).

b) activated carbon in pellet form (500 x magnification).

c) activated carbon in pellet form (5000 x magnification).

Fig. 4.1 SEM images of different samples

analytical capabilities, allowing a more complete understanding of the structure and properties of materials. Some of them can be:

- Atomic force microscopy (AFM): This technique uses an extremely sharp probe to scan the sample surface and provide high-resolution images of topography and mechanical properties at the nanometre scale. AFM is especially useful for studying vacuum or electron sensitive samples and can operate in variety of environments, including liquids
- Magnetic force microscopy (MFM): Like AFM, MFM uses a magnetic field-sensitive probe to study the distribution and magnetic properties of the sample surface at the nanometre level. This technique is valuable for studying magnetic materials and investigating phenomena such as domain magnetization and magnetic interactions.
- Laser scanning confocal microscopy (LSCM): LSCM uses a lase to scan the sample surface and obtain high-resolution three-dimensional images. This technique is especially useful for studying the morphology and three-dimensional structure of biological samples and transparent materials, allowing the observation of surface and deep features in detail.
- Total internal reflection optical microscopy (TIRF): This technique takes advantage of the total internal reflection phenomenon to study the surface of samples that are in contact with a liquid medium, such as living cells or liquid–solid interfaces. TIRF provides high-resolution images of the surface layer of the sample, allowing the study of dynamic processes in real time, such as cell adhesion and molecule–molecule interaction at the liquid–solid interface.
- Fourier transform infrared spectroscopy: Infrared absorption spectroscopy has its origin in molecular vibrations. The infrared spectrum of a molecule is obtained because of measuring the intensity of an absorbed external radiation for each wavelength, which makes possible the transition between two different vibrational energy levels. Each of these characteristic energy absorptions corresponds to a vibrational motion of the atoms in the molecule. Applied to the case of characterizing the surface chemistry of a material, the rationale is the same, the "vibrating" molecules being those on the surface functional groups.

Table 4.1 shows which type of groups the compound contains according to the frequency band in which it is found [9].

Next, in order to know the possible modifications in the surface chemistry after applying the HTC process to a precursor, being PP the precursor and PP-HC the product transformed by HTC, the infrared spectra (FT-IR) of the samples are analysed and the FTIR analysis of the samples is instructed as an example. From Fig. 4.2, indeed, the surface groups of PP are considerably modified when subjected to the HTC process at high temperatures (230 °C). In particular, the loss of some spectral bands corresponding to the C–O bond stress vibration in pyranose ring alcohols (1110, 1060 and 1045 cm^{-1}) typical of cellulose is evident. Likewise, some bands are much better defined. This is the case of the bands recorded at 160–1500 cm^{-1} and at 1200 cm^{-1}, which could be assigned to υ (C=O) vibrations in olefinic and/or aromatic structures and to υ (C–O) vibrations in ether-type structures. Both types of

Table 4.1 FTIR functional group database

Peak	Group	Class	Peak	Group	Class	Peak	Group	Class
3584–3700	O–H stretching	Alcohol	1735–1750	C=O stretching	δ-lactone	1000–1400	C–F stretching	Fluoro compound
3200–3550	O–H stretching	Alcohol	1745	C=O stretching	cyclopentanone	1310–1390	O–H bending	Phenol
3500	N–H stretching	Primary amine	1720–1740	C=O stretching	Aldehyde	1335–1372	S=O stretching	Sulphonate
3300–3400	N–H stretching	Aliphatic primary amine	1715–1730	C=O stretching	α,β-unsaturated ester	1335–1370	S=O stretching	Sulphonamide
3310–3350	N–H stretching	Secondary amine	1705–1725	C=O stretching	Aliphatic ketone	1342–1350	S=O stretching	Sulphonic acid
2500–3300	O–H stretching	Carboxylic acid	1706–1720	C=O stretching	Carboxylic acid	1300–1350	S=O stretching	Sulphone
2700–3200	O–H stretching	Alcohol	1680–1710	C=O stretching	Conjugated acid	1266–1342	C–N stretching	Aromatic amine
2800–3000	N–H stretching	Amine salt	1685–1710	C=O stretching	Conjugated aldehyde	1250–1310	C–O stretching	Aromatic ester
3267–3333	C–H stretching	Alkyne	1690	C=O stretching	Primary amide	1200–1275	C–O stretching	Alkyl aryl ether
3000–3100	C–H stretching	Alkene	1640–1690	C=N stretching	Imine/oxime	1020–1250	C–N stretching	Amine
2840–3000	C–H stretching	Alkane	1666–1685	C=O stretching	Conjugated ketone	1200–1225	C–O stretching	Vinyl ether
2695–2830	C–H stretching	Aldehyde	1680	C=O stretching	Secondary amide	1163–1210	C–O stretching	Ester

(continued)

Table 4.1 (continued)

Peak	Group	Class	Peak	Group	Class	Peak	Group	Class
2550–2600	S–H stretching	Thiol	1680	C=O stretching	Tertiary amide	1124–1205	C–O stretching	Tertiary alcohol
2349	O=C=O stretching	Carbon dioxide	1650	C=O stretching	δ-lactam	1085–1150	C–O stretching	Aliphatic ether
2250–2275	N=C=O stretching	Isocyanate	1668–1678	C=C stretching	Alkene	1087–1124	C–O stretching	Secondary alcohol
2222–2260	C≡N stretching	Nitrile	1665–1675	C=C stretching	Alkene	1050–1085	C–O stretching	Primary alcohol
2190–2260	C≡C stretching	Alkyne	1665–1675	C=C stretching	Alkene	1030–1070	S=O stretching	Sulphoxide
2140–2175	S–C≡N stretching	Thiocyanate	1626–1662	C=C stretching	Alkene	1040–1050	CO–O–CO stretching	Anhydride
2120–2160	N=N=N stretching	Azide	1648–1658	C=C stretching	Alkene	985–995	C=C bending	Alene
2150	C=C=O stretching	Ketene	1600–1650	C=C stretching	Conjugated alkene	960–980	C=C bending	Alkene
2120–2145	N=C=N stretching	Carbodiimide	1580–1650	N–H bending	Amine	885–895	C=C bending	Alkene
2100–2140	C≡C stretching	Alkyne	1566–1650	C=C stretching	Cyclic alkene	550–850	C–Cl stretching	Halo compound
1990–2140	N=C=S stretching	Isothiocyanate	1638–1648	C=C stretching	Alkene	790–840	C=C bending	Alkene
1900–2000	C=C=C stretching	Allene	1610–1620	C=C stretching	α,β-unsaturated ketone	665–730	C=C bending	Aklene

(continued)

Table 4.1 (continued)

Peak	Group	Class	Peak	Group	Class	Peak	Group	Class
2000	C=C=N stretching	Ketenimine	1500–1550	N–O stretching	Nitro compound	515–690	C–Br stretching	Halo compound
1650–2000	C–H bending	Aromatic compound	1465	C–H bending	Alkane	500–600	C–I stretching	Halo compound
1818	C=O stretching	Anhydride	1450	C–H bending	Alkane	860–900	C–H bending	1,2,4-trisubstituted
1785–1815	C=O stretching	Acid halide	1380–1390	C–H bending	Aldehyde	860–900	C–H bending	1,3-disubstituted
1770–1800	C=O stretching	Conjugated acid halide	1380–1385	C–H bending	Alkane	790–830	C–H bending	1,4-disubstituted
1775	C=O stretching	Conjugated anhydride	1395–1440	O–H bending	Carboxylic acid	790–830	C–H bending	1,2,3,4-tetrasubstituted
1770–1780	C=O stretching	vinyl/phenyl ester	1330–1420	O–H bending	Alcohol	760–800	C–H bending	1,2,3-trisubstituted
1760	C=O stretching	Carboxylic acid	1380–1415	S=O stretching	Sulphate	735–775	C–H bending	1,2-disubstituted
1735–1750	C=O stretchin	Esters	1380–1410	S=O stretching	Sulphonyl chloride	730–770	C–H bending	Monosubstituted
						680–720		Benzene derivative

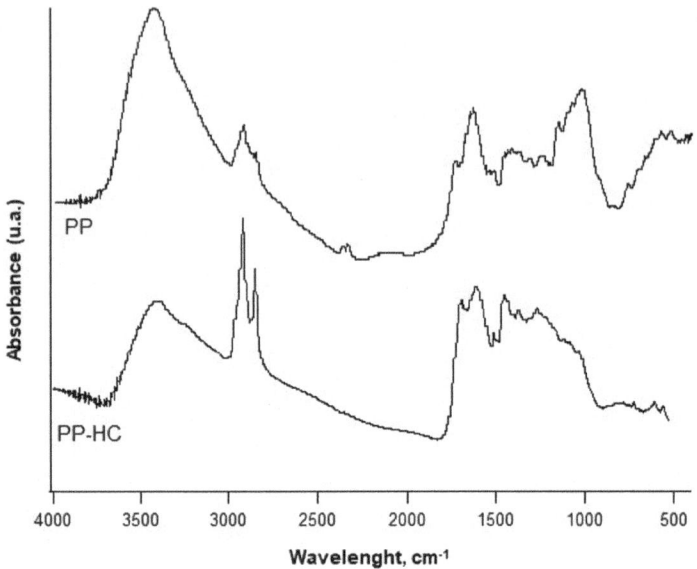

Fig. 4.2 FT-IR spectra for PP and PP-HC

structures would be more resistant to the action of HTC and would therefore merely concentrate in the carbonized product.

Specific surface analysis (BET)

Physical adsorption of gases is one of the most immediate methods to determine a great deal of information about the textural characteristics of adsorbents.

Nitrogen is the most used adsorbate for the characterization of adsorbent solids. This is due to the possibility of covering a wide range of relative pressures, due to the high saturation pressure of this gas, thus allowing to obtain information concerning the micro and mesopore regions. During nitrogen adsorption, the pores of the solid are filled by the adsorbate, a process that begins in the micropores. In this first phase, there are two stages: in the first, called primary filling of the micropores, the adsorbate molecules penetrate the pores with molecular dimensions at very low relative pressures. In the second stage, the filling of the wider micropores occurs, with interaction between the adsorbed molecules. Thereafter, the filling of the mesopores begins, being possible the filling by physical adsorption and by capillary condensation.

On the other hand, the shape of the nitrogen adsorption isotherm can provide very important information and deserves an analysis beyond the simple calculation of the characteristic parameters reporting the extent of adsorption. Brunauer et al. [10] performed a classification of gas adsorption isotherms based on empirical data obtained with porous or non-porous solids. The classification proposed (B.D.D.T.) and currently recommended by International Union of Pure and Applied Chemistry (IUPAC) considers six types of adsorption isotherms (Fig. 4.3).

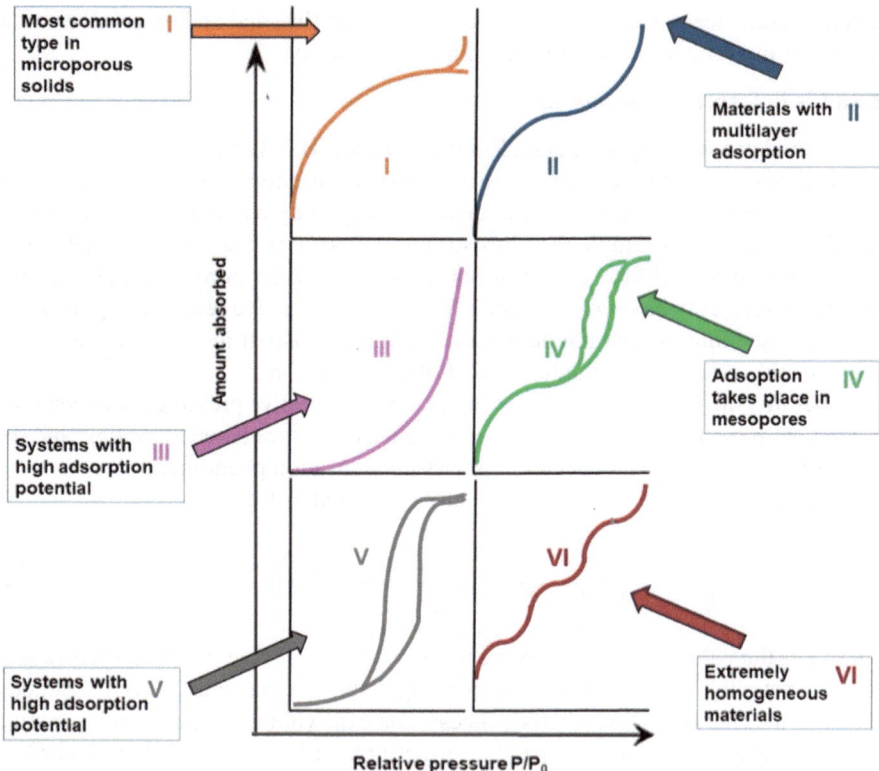

Fig. 4.3 Adsorption isotherm classification proposed by B.D.D.T. Adapted from B.D.D.T [10]

- The isotherms known as type I or Langmuir type are characteristic of essentially microporous solids, where the volume of nitrogen adsorbed takes place mainly at low values of relative pressures close to 0.05; whereas, at higher relative pressures, the amount adsorbed remains constant or increases slightly.
- Type II isotherms show an inflection around relative adsorption pressures of 0.1 and 0.9. These isotherms are characteristic of materials where multilayer adsorption is present.
- Type III isotherms of the B.D.D.T. are convex, ascending and characteristic of systems with high adsorption potential.
- Type IV isotherms are like type II, but adsorption takes place additionally in the mesopores present.
- Type V isotherms are characteristic of adsorption in systems with low energy potentials on very mesoporous homogeneous surfaces, and type VI isotherms correspond to extremely homogeneous materials.

The following is a brief review of the different models and methods proposed in the literature for the determination, from nitrogen adsorption data, of characteristic textural parameters such as the specific surface area (internal and external) and the volume of micro- and mesopores of the adsorbent solids.

Model B.E.T. specific surface area

The method proposed by Brunauer, Emmett y Teller (B.E.T.) [10] can be considered as an extension of Langmuir's theory to multilayer adsorption. These authors consider multilayer adsorption, admitting that there is always a dynamic equilibrium in each adsorbed layer, and resemble the physisorption of N_2 in porous solids to that which takes place on the surface of a non-porous solid. Furthermore, consider that the adsorption surface to be energetically uniform, so that all active centres are equivalent. Finally, they point out that the heats of adsorption are equal in all layers above the first one and coincide with the latent heat of condensation.

Generally, the model is applicable in the range of relative pressures between 0.05 and 0.2, with the lower limit due to the existence of surface heterogeneities, and the upper limit due to the existence of capillary condensation phenomena.

The B.E.T. Equation (Eq. 4.4) is usually expressed as follows:

$$\frac{P}{V(P_0 - P)} = \frac{1}{V_m} + \frac{(C - 1)^{P/P_0}}{V_m C}, \tag{4.4}$$

where V is the volume of gas at relative pressure P/P_0, P_0 is the saturation pressure of the adsorbate, V_m is the volume corresponding to the monolayer and C is a parameter related to the energy involved in the process. Once the volume of the monolayer (V_m) is known, the calculation of the specific surface area ($m^2 \, g^{-1}$) is carried out according to the following equation (Eq. 4.5):

$$S_{BET} = \frac{V_m N_A A_m}{22414}, \tag{4.5}$$

where N_A is the Avogadro number and A_m is the area occupied by the adsorbed molecule (N_2) at the working temperature (77 K) on the surface of the solid (0.162 nm^2).

Dubinin-Radushkevich model micropore volume

The Dubinin-Radushkevich (D.R.) [11] model is based on Polanyi's potential theory. According to Dubinin, the filling of the micropores occurs with the adsorbate liquid state in a physical adsorption process, instead of the formation of successive layers of adsorbate on the surface of the walls, as assumed in the B.E.T. theory. The D.R. equation (Eq. 4.6) is usually written:

$$\log V = \log V_0 - D \log^2(P_0/P), \tag{4.6}$$

where V is the adsorbed volume at pressure P, V_0 is the total micropore volume and D is a constant.

The applicability range of the D.R. oscillates between values of relative pressures between 10^{-5} and 0.4 for activated carbons, a lower limit due to effects produced in the ultra-micropores and higher limit because the filling of the mesopores begins at these pressures, both aspects not contemplated in the theoretical approaches. In fact, the deviations found in the application of the D.R. equation may be due to several factors such as:

- The distribution of adsorption potentials is not adjustable to a Gaussian function.
- The existence of a phase change of the adsorbate during the adsorption process.
- Due to several ranges of microporosity distribution lead to almost linear D.R. representations over the whole range of relative pressures. However, the widening of the micropores leads to a smaller range of compliance with the D.R. equation, with a second section appearing at higher relative pressures, and the representation becoming more curved as the mean pore size increases [12].

Method of Gregg and Sing mesopore volume

According to Sing [13], whose work was based on the fulfilment method of Gurvistch, it can be considered that the volume of micropores is approximately equal to the adsorbed volume, expressed as liquid at relative pressures of 0.1. It is also considered that at values of relative pressures close to unity (around 0.95), filling of both micropores and mesopores occurs. Thus, the mesopores volume can be estimated as the difference between the two parameters above.

Sing's method α_s external surface

Due to the complexity of the physisorption phenomenon, it is useful to use empirical procedures for isotherm analysis. Among them, models using standard adsorption isotherms of a non-porous reference material are recommended. These methods allow the evaluation of both the non-microporous surface area of the carbon and the micropore volume. Among these methods, Sing's method [14, 15] is considered, which contemplates that the adsorption data must be normalized to a given value of relative pressure and thus defines the parameter α_s as the quantity adsorbed at 0.4, the relative pressure at which nitrogen adsorption has already ended in the micropores and there are no capillary condensation phenomena.

Sing's method α_s consists, first, in constructing the reference isotherm in a reduced form, representing the parameter α_s versus the corresponding values of relative pressures. Thus, the α_s curve of the material to be analysed is constructed, representing the values of the absorbed volume of nitrogen of the material to be analysed (V_{sample}) versus the α_s parameter of the reference material corresponding to the same relative pressure. This representation takes linear form in different ranges of α_s values; the corresponding ordinate in the origin being equal to the volume of the micropores. The external area of the sample to be analysed (S_{EXT}) can be determined from the surface area of the reference solid (S_{ref}), by means of the equation (Eq. 4.7):

$$S_{EXT} = X_\alpha \left(\frac{S_{ref}}{V_{(ref)}P/P_0 = 0.4} \right), \tag{4.7}$$

where $X\alpha$ is the value of the slope of the linear part of the corresponding representation, and $V_{(ref)}P/P_0 = 0.4$ is the volume adsorbed by the reference solid at the relative pressure of 0.4. The non-porous reference material used must have a surface chemistry similar to that of the material under analysis. In this sense, there are several charcoal standards proposed in the literature, among which the one proposed by Rodríguez-Reinoso et al. [16] from olive stones and Carrot y et al. (from a non-graphitized charcoal) stand out [17].

An example of N_2 adsorption isotherm at 77 K of an unknown activated carbon is shown below (Fig. 4.4). It can be seen that it is typical of the type I of B.D.D.T. classification [10], characteristic of microporous materials. It is observed that the adsorption process takes place very strongly at low relative pressures and then, once the micropores are filled with adsorbate, the amount adsorbed is approximately constant, so that there is a well-defined plateau. By application to the N_2 adsorption data at 77 K of the B.E.T. model (Eqs. 4.4 and 4.5), Dubinin- Radushkevich (Eq. 4.6) and the method of Greeg and Sing [13] modified according to the variant of Reinoso et al. [18] (taking V_{mi} as the one determined from D.R.), the value of the specific surface area BET (S_{BET}), micropore volume (V_{mi}) and mesopore volume (V_{me}), respectively, were determined. Similarly, by applying the α_s method (Eq. 4.7), the external surface area, S_{EXT} was evaluated. The value corresponding to the percentage of internal surface area, S_{INT}, was calculated assuming $SI_{NT} = S_{BET} - S_{EXT}$. The values obtained are tabulated in Table 4.2.

As can be seen in Table 4.2, the activated carbon has a high specific surface area and micropore volume and almost negligible mesoporosity development.

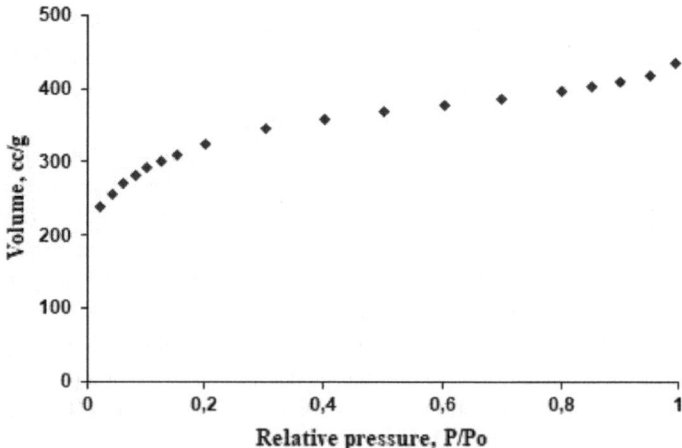

Fig. 4.4 N_2 adsorption isotherm at 77 K for activated carbon

Table 4.2 Textural characteristics of activated carbon

Adsorption of N_2 a 77 K					Porosimetry	
S_{BET} (m^2 g^{-1})	$V_{mi(DR)}$ (cm^3 g^{-1})	V_{me} (cm^3 g^{-1})	S_{EXT} (m^2 g^{-1})	S_{INT} (%)	V_{meP} (cm^3 g^{-1})	V_{maP} (cm^3 g^{-1})
930	0.490	0.063	78	92	0.121	0.301

Mercury Porosimetry Meso and Macropore Volume

This technique is based on the fact that when a solid is immersed in mercury, the penetration of this fluid into the pores of the solid does not occur unless external pressure is applied; the higher the pressure administered, the greater the intrusion. Based on a porosity model that assumes cylindrical, non-intersecting pores, Washburn proposed the following equation (Eq. 4.8), which relates the pressures applied when mercury is introduced ($\Delta\rho$) to the pore radius (r):

$$r = 2 \cdot \gamma \cos \frac{\theta}{\Delta\rho}, \tag{4.8}$$

where s is the surface tension of mercury (0.480 N m^{-1} at 20 °C) and θ is the contact angle of this liquid with the surface of the solid. This value is usually taken to be equal to 140°.

From the mercury intrusion data, it is possible to determine the macropore volume, as the intrusion volume corresponding to the 50 nm pore diameter, and the mesopore volume, as the difference between the intrusion volume at 2 nm diameter and the previous one (sometimes the sensitivity of the porosimeter does not allow to reach pore sizes of 2 nm, taking in this case, the lowest measurement provided by the equipment).

Since meso- and macropores are generally not uniform in shape (nor cylindrical in their entirety), the results determined from mercury porosimetry have to be taken with certain reservations. When applying the Washburn equation to the mercury intrusion curve, it is assumed that the contact angle remains unchanged and that the porous structure does not undergo any irreversible change or compress when subjected to high pressures, which may occur, depending on the physical properties of the material. Even so, this technique is extremely useful and, when compared in a complementary way with other techniques, often leads to reasonably similar results.

Figure 4.5 shows, as an example, the mercury intrusion curves and the corresponding pore size distribution of an unknown activated carbon. From the experimental data, the meso- and macropore volumes were found (V_{meP} and V_{maP}), emissions, which are listed in Table 4.2. The volume of mercury intrusion increases gradually over the entire pore size range studied, with the most marked growth for the smallest pore diameters.

Transmission electron microscopy (TEM)

TEM is a high-resolution microscopy technique that allows the internal structure of materials to be observed on a nanometre scale. In the case of HCs, TEM can reveal the presence of nanostructures, such as nanoparticles, nanotubes or nanopores, as well as the distribution of elements at the atomic level. This technique is particularly useful to characterize the morphology and nanostructure of HCs, as well as to study the influence of synthesis conditions on the formation of these structures. The information obtained by TEM is essential to understand the relationship between the nanoscale structure and the macroscopic properties of HCs.

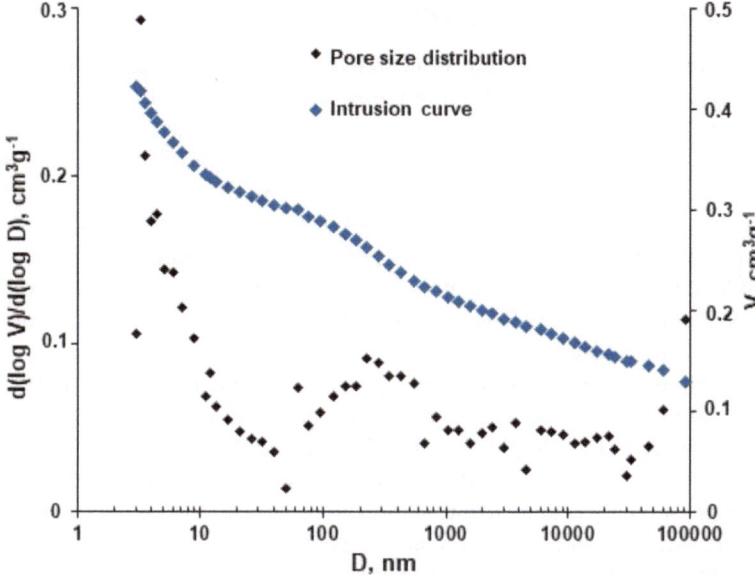

Fig. 4.5 Hg intrusion curves and corresponding activated carbon pore size distribution

Nuclear magnetic resonance spectroscopy (NMR)

NMR is an analytical technique for studying the molecular structure and chemical composition of materials, including HCs. In the case of HCs, NMR can be used to analyse carbon connectivity, identify functional groups and determine the distribution of chemical species in the sample. NMR signals provide information on the presence of different carbon atoms in the molecular structure of HCs and their chemical environment. This technique is especially useful for studying the chemical composition and molecular structure of HCs, as well as for investigating the chemical changes that occur during their synthesis and processing.

X-ray excited photoelectron spectroscopy (XPS)

XPS is a surface analysis technique that allows the elemental composition and surface chemical structure of materials to be studied. In the case of HCs, XPS can be used to analyse the surface composition, identify the functional groups present and characterize the surface chemistry. XPS measurements provide information on the distribution of elements and oxidation states on the surface of HCs, which is important for understanding their chemical reactivity, their affinity for certain contaminants and their interaction with other materials in specific applications. This technique is particularly useful for studying the surface chemistry of modified HCs and for investigating adsorption and catalysis mechanisms in heterogeneous systems.

Thermogravimetric analysis (TGA/DTG)

Thermogravimetric analysis (TGA) is an analytical technique used to study changes in the mass of a sample as a function of temperature or time, providing information on its thermal stability and composition. In this technique, the sample is gradually heated in a controlled environment (usually inert, such as nitrogen) to about 800 °C while its mass is continuously measured. Changes in the mass of the sample are recorded as a function of temperature or time. Normally between 25 and 150 °C, the dehydration reaction occurs, then there is a stage around 150–350 °C where hemicellulose is broken down and there is a significant loss in mass, afterwards the same is true with the cellulose between 350 and 450 °C and finally lignin decomposes until the end of the heating process.

Differential thermogravimetric analysis (DTG) is an extension of TGA that involves the derivative of the thermogravimetry curve to obtain additional information about changes in the sample. The first derivative of the thermogravimetry curve is calculated to obtain the rate of mass change with respect to temperature or time, which allows for more accurate identification and characterization of decomposition, oxidation, desorption, and other thermal events occurring in the sample.

An example of a thermogravimetric analysis for an unknown coal sample is shown in Fig. 4.6. The blue line represents the TGA, and the green line represents the DTG. If TGA is considered, it can be observed how this sample loses mass around 350 °C and then reaches 0. In the DTG curve, it can be seen that, at approximately 400 °C, the change of slope occurs.

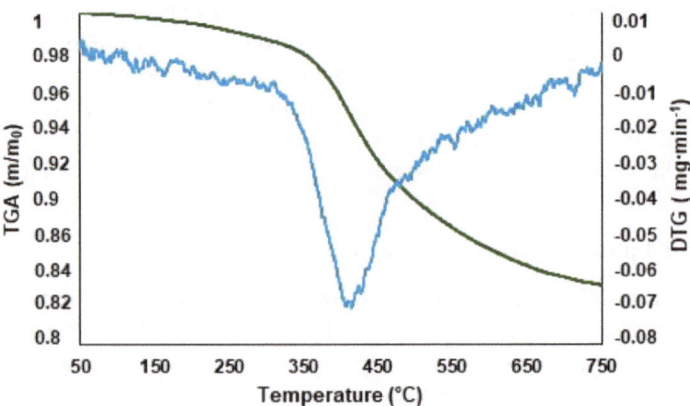

Fig. 4.6 Thermogravimetric analysis for an unknown coal sample

4.1.3   Characterization of PW

The most commonly used characterization techniques for the analysis of PW are presented below.

pH and electrical conductivity (EC)

The pH is an index that measures the acidity or alkalinity of an aqueous solution, in this case PW. It is measured using a pH metre, an instrument consisting of a pH-sensitive electrode that analyses the activity of hydrogen ion in aqueous solutions, indicating their degree of acidity or alkalinity expressed as pH.

Electrical conductivity (EC) reveals the ability of the PW to conduct electric current using a conductivity metre consisting of a conductivity probe.

Total organic carbon (TOC)

Total organic carbon (TOC) is the amount of total carbon bound to an organic compound and is used, in this case, to measure the quality of PW as a non-specific indicator. It is measured using a TOC analyser which indicates the amount of carbon dioxide that is generated when oxidizing organic matter under special conditions.

Chemical Oxygen Demand (COD)

The chemical oxygen demand with respect to the PW is the amount of oxygen required to oxidize the matter present in the sample (organic and inorganic). A chemical oxidant such as potassium dichromate in an acid medium is generally used to oxidize the sample. The COD is expressed in milligrams of diatomic oxygen per litre (mg O_2/L), and although the main contributor to the chemical oxygen demand is the dissolved or suspended organic matter, some inorganic components present in the PW (sulphides, sulphites, iodides, etc.) also contribute to the result of the COD and are reflected in the measurement.

Biochemical oxygen demand measurement (BOD5)

The biochemical oxygen demand (BOD5) of the PW represents the amount of oxygen consumed by the aerobic microorganisms present in the PW when degrading organic matter. It can therefore be considered as a measure of the biodegradability of a water under aerobic conditions. The incubation period considered most suitable for this measure is five days, BOD5.

4.2   Preventing Hazards in the Practice of HTC

As is well known, the HTC technique involves the handling of equipment requiring high temperatures and pressures, which can imply various hazards if certain precautions are not taken. Safety in laboratories must be a priority for all users, therefore it is necessary that the rules are properly understood and applied in order to achieve

Table 4.3 Risks of the HTC

Hazards	Activities
Thermal	High-temperature reactor handling
Pressure	Handling of reactors operating at high pressure
Chemical	Use of potentially hazardous reagents and products
Mechanical	Use of equipment such as crushers and grinders

a safe working environment. The hazards to which the HTC process requires most attention are presented in Table 4.3 according to the main activities carried out in HTC:

Therefore, in order to avoid any incidence of the risks presented, it is necessary to receive adequate training on the use of the equipment and procedures, as well as to thoroughly read the protocols and manuals of each piece of equipment to be used. In addition, it is highly advisable to bear in mind the following:

- Use of personal protective equipment: heat-resistant thermal gloves, as well as disposable gloves that protect against chemical products; goggles to protect against splashes; lab coats to protect skin and clothing; ear protection when working with equipment such as crushers, sieving machines…
- When working with pressure reactors, it is necessary to check them beforehand to make sure that there is no damage or wear on each of their components.

Always work within the stipulated design conditions; never exceed the temperature and pressure recommended by the manufacturer.

Before starting an HTC reaction, make sure that the reactor and its valves are tightly closed and wait until the reaction is completely cooled down before handling.

- As for process waters or chemicals that may be used during HTC, store them in appropriate and properly labelled locations.
- Before starting work in the laboratory, be aware of the emergency plan, including emergency exits and assembly points. It is advisable to be familiar with the use of equipment such as fire extinguishers and safety showers.
- It is advisable to have a basic knowledge of first aid and how to act in case of burns, cuts or chemical exposure and to have an accessible and well-stocked first aid kit.

4.3   Sustainability and Environmental Impact at HTC

The use of biomass or other waste as sustainable source of energy carriers and to produce value-added products is a proven alternative that has led to the emergence of the biorefinery concept. The advantages of using these sources are well known, but the suitability of the process will depend on the energy and economic costs and also on the environmental impacts associated with the techniques. In this context, finding more environmentally processes to convert these materials into valuable products, which minimize the potential risks to the environment, is essential to make the whole concept responsible. HTC is therefore a cost-effective, promising and environmentally friendly alternative to traditional thermochemical processes to convert biomass into high value-added materials, with a higher energy content compared to the original material, contributing to the circular economy by reducing dependence on fossil resources, minimizing waste and acting as a carbon sink by contributing to its long-term sequestration.

This technique, HTC, compared to other traditional thermochemical processes, such as combustion or pyrolysis, significantly improves the reduction of greenhouse gas emissions (GHG) since it converts wet biomass or biomass mixed with water into HC and PW, encapsulating the carbon present in the biomass and preventing its release as CO_2. It should be noted at this point that HTC allows the use of agricultural, forestry and municipal waste with high moisture content that would otherwise be discarded or incinerated, thus reducing the amount of solid waste.

As discussed in Chap. 1, the HC produced has multiple applications and can be used as a solid fuel instead of fossil coal, thus reducing the carbon footprint of energy production. It can also be used in wastewater treatment, acting as an adsorption medium for organic and inorganic pollutants.

In addition to the solid product HC, the HTC process generates PW that can be treated a reused, and in some cases, valuable by-products such as organic acids and nutrients can also be recovered, which can in turn be used as liquid fertilizers.

In terms of energy consumption, HTC is relatively efficient due to the moderate temperature conditions (180–250 °C) compared to pyrolysis (600–800 °C). Although energy is required to heat the reactor at milder temperatures, it could be fuelled by renewable sources, as well as incorporating waste heat recovery systems can further improve the energy efficiency of the HTC process.

All in all, HTC fosters, on the one hand, innovation in sustainable technologies and the development of new, more environmentally friendly industrial processes and, on the other hand, economic and social sustainability by creating valuable products from waste that can generate new economic and employment opportunities in rural and urban sectors, contributing to sustainable development and economic resilience.

4.4 Exercises

1. Fifty grams of a dry biomass are weighed to carry out an HTC process. For this purpose, 100 g of water are added and introduced into the reactor where the reaction will be carried out. The biomass is initially dry. At the end of the process, 28 g of dried HC are obtained as a solid phase and 15 g of PW as a liquid phase. The gas phase is not accounted for. Calculate the yield by weight for each product phase obtained in the HTC process. Provide the detailed calculations and the results obtained.

2. Fourier transform infrared spectroscopy (FTIR) analysis has been performed on an unknown sample, the graph of which is shown below. You are asked to observe and identify the characteristic peaks of interest in the FTIR spectrum (between 4000 and 400 cm^{-1}) to determine the functional groups present in the sample. For this, you can make use of a reference frequency table for functional groups.

3. The weight loss plot of a biomass sample as a function of temperature obtained by thermogravimetric analysis (TGA) is presented below. The sample was heated from 25 to 800 °C with a heating ramp of 10 °C/min in an inert atmosphere (N_2). From the graph below:

    (a) Identify the main stages of thermal decomposition of the sample.
    (b) Determine the temperature at which the maximum weight loss (DTG peak) is observed.

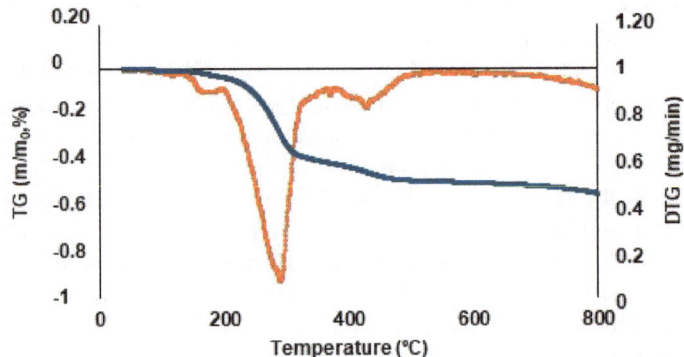

4.5 Questions

1. Why is it sometimes necessary to grind the precursor biomass before the HTC process?

    (a) To increase the moisture content
    (b) To reduce the particle size and thereby improve the homogeneity and efficiency of the process
    (c) To increase the separation of the liquid and solid phases
    (d) To be able to measure the PCZ of the sample

2. Why is it necessary to use a pressure reactor in the HTC process?

    (a) Because the pH of the sample needs to be measured during the process
    (b) Because the sample has to be heated above 500 °C
    (c) Because it is necessary to maintain the solid and liquid water phases under controlled conditions of temperature and pressure
    (d) Crushing the biomass

3. Which laboratory apparatus is not necessary to carry out the HTC technique?

   (a) Pressure reactor
   (b) Heating oven
   (c) Balance
   (d) Spectrophotometer

4. By calculating the higher heating value, what value is obtained?

   (a) The amount of water in the sample
   (b) The amount of heat released during complete combustion per unit mass
   (c) The percentage of carbon in the sample
   (d) The ash in the sample when burned

5. What is the name given to the analysis used to determine the C, H and N content of a biomass or HC sample?

   (a) Infrared spectroscopy
   (b) Elemental analysis
   (c) Gas chromatography
   (d) Immediate analysis

6. During elemental analysis, how is the percentage of O in the sample determined?

   (a) By titration
   (b) By difference with the total C, H, N, and S content
   (c) By difference from the immediate analysis
   (d) With a calorimeter

7. What is the formula used in the FC fixed carbon analysis?

   (a) $FC = [100 - (Moisture + Ash + Volatile\ Matter)]$
   (b) $FC = [100 + (Moisture + Ash + Volatile\ Matter)]$
   (c) $FC = [100 - (Ash + Volatile\ Matter)]$
   (d) $FC = [100 - (Moisture + Ash)]$

8. Which methodology can be used to determine the cellulose, hemicellulose and lignin content in a solid sample?

   (a) Gas chromatography
   (b) Fibre analysis
   (c) Absorption spectroscopy
   (d) Immediate analysis

9. Which instrument is used to measure the acidity or basicity number of the process water (PW) after the HTC reaction?

   (a) Spectrophotometer
   (b) pH metre
   (c) Calorimeter
   (d) Chromatograph

10. Which analytical technique allows visualization of the morphology and surface structure of the materials obtained from the HTC process?

    (a) FTIR
    (b) Scanning electron microscopy (SEM)
    (c) Elemental analysis
    (d) Calorimetry

11. To which materials can the scanning electron microscopy (SEM) technique be applied?

    (a) Only for organic materials
    (b) For organic and inorganic materials
    (c) Only for inorganic materials
    (d) None of the above is correct

12. Which of the following characteristics cannot be directly observed by SEM analysis?

    (a) Roughness
    (b) Chemical composition
    (c) Porosity
    (d) Texture

13. Which of the following functional groups could be identified in a sample by FTIR in the 3200–3550 cm^{-1} region?

    (a) C–H stretching (alkane)
    (b) N–H stretching (amine salt)
    (c) O–H stretching (alcohol)
    (d) C≡N stretching (nitrile)

14. Which gas is normally used in the BET method to characterize adsorbent solids?

    (a) Oxygen
    (b) Argon
    (c) Nitrogen
    (d) Carbon dioxide

15. What types of pores are filled first during nitrogen adsorption in BET analysis?

    (a) Macropores
    (b) Mesopores
    (c) Micropores
    (d) Nanopores

16. Type I or Langmuir adsorption isotherms are characteristic of solids that are essentially....

    (a) Macroporous
    (b) Mesoporous

   (c) Microporous
   (d) Nanoporous

Competing Interests The authors have no conflicts of interest to declare that are relevant to the content of this chapter.

References

1. C.H. Dang, G. Cappai, J.W. Chung, C. Jeong, B. Kulli, F. Marchelli, K.S. Ro, S. Román, Research needs and pathways to advance hydrothermal carbonization technology. Agronomy **14**(2), 247 (2024)
2. X. Chen, J. Zhang, Q. Lin, G. Li, X. Zhao, Dispose of Chinese cabbage waste via hydrothermal carbonization: hydrochar characterization and its potential as a soil amendment. Environ. Sci. Pollut. Res. Int. **30**(2), 4592–4602 (2023). https://doi.org/10.1007/s11356-022-22359-4
3. M. Yan, Y. Liu, Y. Song, A. Xu, G. Zhu, J. Jiang, D. Hantoko, Comprehensive experimental study on energy conversion of household kitchen waste via integrated hydrothermal carbonization and supercritical water gasification. Energy **242**, 123054 (2022). https://doi.org/10.1016/j.energy.2021.123054
4. J. Fang, L. Zhan, Y.S. Ok, B. Gao, Minireview of potential applications of hydrochar derived from hydrothermal carbonization of biomass. J. Ind. Eng. Chem. **57**, 15–21 (2018). https://doi.org/10.1016/j.jiec.2017.08.026
5. M. He, X. Zhu, S. Dutta, S.K. Khanal, K.T. Lee, O. Masek, D.C. Tsang, Catalytic co-hydrothermal carbonization of food waste digestate and yard waste for energy application and nutrient recovery. Biores. Technol. **344**, 126395 (2022). https://doi.org/10.1016/j.biortech.2021.126395
6. G. Mannarino, A. Sarrion, E. Diaz, R. Gori, M.A. De la Rubia, A.F. Mohedano, Improved energy recovery from food waste through hydrothermal carbonization and anaerobic digestion. Waste Manage. **142**, 9–18 (2022). https://doi.org/10.1016/j.wasman.2022.02.003
7. Y. Wei, S. Fakudze, Y. Zhang, R. Ma, Q. Shang, J. Chen, C. Liu, Q. Chu, Co-hydrothermal carbonization of pomelo peel and PVC for production of hydrochar pellets with enhanced fuel properties and dechlorination. Energy**239**, 122350 (2022). https://doi.org/10.1016/j.energy.2021.122350
8. D. Díez, A. Urueña, R. Piñero, A. Barrio, T. Tamminen, Determination of hemicellulose, cellulose, and lignin content in different types of biomasses by thermogravimetric analysis and pseudocomponent kinetic model (TGA-PKM method). Processes **8**(9), 1048 (2020). https://doi.org/10.3390/pr8091048
9. FTIR Functional Group Database Table with Search—InstaNANO. https://instanano.com/all/characterization/ftir/ftir-functional-group-search/. Accessed 27 June 2024
10. S. Brunauer, L.S. Deming, W.E. Deming, E. Teller, On a theory of the van der Waals adsorption of gases. J. Am. Chem. Soc. **62**(7), 1723–1732 (1940)
11. M.M. Dubinin, in *Progress in Surface and Membrane*, ed. by D.A. Candenhead (Science 9, Academic Press, London, 1975)
12. J. Garrido, A. Linares-Solano, J.M. Martin-Martinez, M. Molina-Sabio, F. Rodriguez- Reinoso, R. Torregrosa, Use of nitrogen vs. carbon dioxide in the characterization of activated carbons. Langmuir **3**(1), 76–81 (1987)
13. S.J. Gregg, K.S.W. Sing, *Adsorption, Surface Area and*, Porosity. (Academic Press, London, 1982)

14. K.S.W. Sing, D.H. Everett, R.A.W. Haul, L. Moscou, R.A. Pierotti, T. Siemieinewska, Reporting physisorption data for gas/solid systems with special reference to the determination of surface area and porosity (Recommendations 1984). Pure Appl. Chem. **57**, 603–609 (1985)
15. K.S.W. Sing, D.H. Everett, R.H. Ottewill (eds.), *Surface Area Determination* (Butterworths, London, 1970)
16. F. Rodriguez-Reinoso, J.M. Martin-Martinez, C. Prado-Burguete, B. McEnaney, A standard adsorption isotherm for the characterization of activated carbons. J. Phys. Chem. **91**(3), 515–516 (1987)
17. P.J.M. Carrott, R.A. Roberts, K.S.W. Sing, Standard nitrogen adsorption data for nonporous carbons. Carbon **25**(6), 769–770 (1987)
18. F. Rodríguez-Reinoso, *Carbón Activado: Estructura, preparación y aplicaciones* (Publicaciones de la Universidad de Alicante, 2006)

Chapter 5
Technical Issues

5.1 Current Technology

At this point, the advantages of hydrothermal and co-hydrothermal carbonization have been proven. Thus, this technology has a promising future, especially applied to green technologies and circular economy, taking part in a biorefinery context. However, as previously explained, this process is relatively innovative, and its implementation at industrial scale (where effectiveness and adaptation to specific industrial circumstances are essential) is, in some cases, still a challenge that research and development can solve.

In that sense, the role of multidisciplinary teams (where professionals devoted to engineering, chemical engineering, environmental sciences, etc., have a great deal to say) is vital in order to make the industrial adaptation of HTC a reality. As explained in Chap. 1, the multidisciplinary nature of HTC could be an interesting starting point for its development at industrial level.

How do we assess the technological maturity of a certain technology to know the room for improvement? For that purpose, there is an interesting scale called technology readiness levels (TRL), developed by NASA (research paradigm in applied science), that classifies a technology according to its degree of applicability [1]. As observed in Table 5.1, there are 9 TRL (which are adapted from aerospace stages to general research), from the earliest stage of development (TRL 1) to a completely mature technology (TRL 9).

Consequently, the TRL evolution of a certain technology could determines, as in the case of the number of published articles or patents about a specific field, the real interest at industrial level and the subsequent feasibility of implementation.

Thus, research and scale-up work is essential to make a certain technology feasible in our day to day, with plenty of multidisciplinary teams devoted to the improvement of the abovementioned stages of development as soon as possible. How is it fundamentally carried out? In essence, scaling-up a technology is based on the main results found at laboratory scale, trying to consider the dimensional ratios (in reactors,

B. Ledesma Cano et al., *Introduction to Hydrocarbonization*,
SpringerBriefs in Applied Sciences and Technology,
https://doi.org/10.1007/978-3-031-70039-2_5

Table 5.1 Different technology readiness levels and their meaning [1]

TRL	Meaning
1	Basic principles have been observed and reported. The beginning of scientific research
2	A technology concept or application has been formulated. Some practical applications have been found, and the possibility of an industrial scale implementation begins to take shape
3	Experiments and analysis (mainly at laboratory level) about feasibility or construction of a proof-of-concept model
4	Test of multiple components at laboratory level
5	Test of multiple components in relevant environments. It implies more rigorous testing than the previous level
6	A fully functional prototype or representational model is obtained, with a demonstration in a relevant environment
7	Demonstration of the prototype or model in a real environment
8	The system is complete, ready for the real implementation into an existing technology or technology system
9	The system is successfully proven in a real environment

for example), kinetics of the process or forces such as pressure. For that purpose, a perfect knowledge of equipment characteristics at industrial scale is essential, as well as selecting the right parameters that can be similarly controlled after scaling-up.

In this sense, there are numerous research works devoted to the real implementation of this technology at industry level, focused on various aspects related to different wastes and their adaptability to HTC. In other words, there is a significant share of research whose main objective is to improve this ranking, including more than 500 patents in a short period of time (since 1996).

Considering the main facts observed in previous chapters, where the main wastes, chemical conditions and products have thoroughly been studied, the potential of this technology has been proven, offering versatile solutions to different environmental and sustainable challenges in business and markets, which seems to gain more and more relevance in the present global scenario.

Indeed, there are several companies that, directly or indirectly, have production lines related to HTC. As observed in Fig. 5.1, the majority of HTC plants are located in Europe (both pilot or demonstration plants and full-scale plants), with nearly 20 plants, followed by Asia with 10 plants [2]. Specifically, Germany presents a high level of specialization, with 6 plants and several companies. These data reveal that HTC is an emerging reality in many developed areas, which indicates the potential of this technology in these regions as well as in the case of developing countries in the long run, where similar wastes can be perfectly managed through the same technology. In fact, in parallel, there are other plants, like wastewater plants, whose increase is considerable both in developing and developed areas [3], where HTC could play an important role, proving good prospects in future. Thus, according to

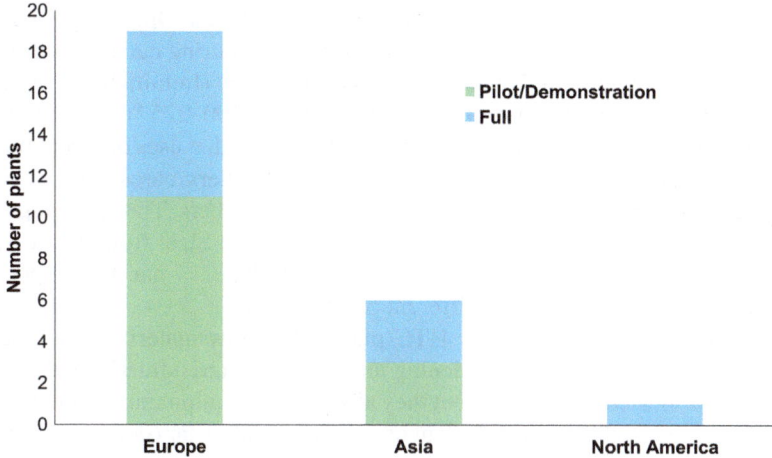

Fig. 5.1 HTC plants ranked by continent (based on [2])

the size of these plants, it can be concluded that this technology has reached TRL9, ranging the technology maturity from TRL7 to TRL9 in the abovementioned cases.

In these plants, there are clear differences, mainly based on the size of the plant and the kind of waste and its availability, which clearly have a strong influence on the kind of reactor used (batch or continuous), the reactor size (from 2 to 3000 L) or plant capacity (up to 200 t/day). In any case, the main aim of these plants is waste management and valorization, with up to 80% of companies devoted to that purpose, whereas other different uses were also found, like carbon removal or chemical production, among others.

To sum up, some of the most important companies that use HTC or similar techniques around the world are included [2]:

- Antaco, Guildford, UK (2011). This is an HTC pilot plant, which was the first one devoted to this technique in the UK. It uses organic wastes derived from food, sludge, agricultural waste, manure, etc. It operates continuously, with the following working conditions: $T = 200$ °C, $P = 25$ bar, $t = 4$–10 h.
- AVA Biochem, Zug, Switzerland (2014). Another HTC plant (in this case, for demonstration). It utilizes C6 sugars from different streams. Demonstration plant for production of 5-HMF (up to 30 tonnes of 100% HMF). It uses a patented water-based process called COBRIS™.
- Calpech, Alicante, Spain (2021). It is an HTC pilot plant that transforms industrial wastes, with a reactor capacity of 50 L.
- Carbensate, Stuttgart, Germany (2024). This is a prototype for an HTC plant. It uses sewage sludge and animal, food and industrial wastes.
- C-Green AB, Solna, Sweden (2014). It uses lignocellulosic biomass such as sewage, cellulose or paper sludge. It has a demonstration plant located in Heinola,

Finland (2020). Reactor size: 5 m^3; Capacity: 18.000 t/year. Operating conditions: Continuous, $T = 200$ °C, $P = 20$ bar, $t = 1$ h, not requiring catalyst addition.

- CPL Industries, Sheffield, UK. Another HTC pilot plant (Imminghan, UK, 2018). Partnership with Ingelia. Operating conditions: $T = 200–225$ °C, $P = 20$ bar.
- Da Invent Co., Ltd., Nagoya, Japan (1992). HTC plant that uses different biomass and wastes such as municipal solid waste (MSW), feathers, carcass, oil extraction residue, agriculture residues or food residues, among others. The operating conditions are the following: Pilot and Full Plant Reactor(s) Size: from 2 L to 15 m^3, capacity: 200 t/day. $T = 200–240$ °C, $P = 20–30$ bar, $t =$ n.a. It is expected to expand its production through different plant projects.
- Ingelia, Valencia, Spain (2010). HTC plant. Main raw materials used: sewage sludge, MSW, plant wastes (gardening and pruning), agricultural, food residues. The plant is located in Valencia, but they also provide equipment for other companies. Operating conditions: Full plant capacity, 14,000 t/year (outputs: 750 tonnes of fertilizer concentrate and 3500 tonnes of bio-coal), the plant operates continuously (claimed to be the first industrial plant to operate in this mode), $T = 180–200$ °C, $P =$ n.a., $t = 4–12$ h. It uses acid as a catalyst.
- SunCoal Industries GmbH, Ludwigsfelde, Germany (2007). HTC and HTT pilot plant. It uses lignocellulosic biomass wastes. The main operating conditions are not specified.

As a conclusion, the existence of these pilot and full plants around the world demonstrates the versatility of HTC, which can be perfectly adapted to different wastes and production rates. In this sense, the role of scaling-up and design is essential, requiring a precise sizing of the equipment depending on different factors.

5.2   General Considerations that Influence the Implementation of HTC Plants

As it has been pointed out in previous chapters of this book, the operating conditions, kinetics, reactions taking place during HTC and the different products obtained throughout the process (with their corresponding properties) have been extensively studied over the years. However, when it comes to the implementation of HTC technology at industrial scale, through real and operational plants, several technical and design challenges arise, which should be pointed out. Without going into technical detail, the following issues should be mentioned.

5.2.1   Process Water and Its Reuse

As previously mentioned, water is one of the main components in HTC process. However, the immediate and main drawback is access to water or, at least, keeping water supply conditions for operation in HTC plants. A possible solution could be water recirculation, consisting of reusing process water after HTC to carry out successive HTC reactions. Several studies have found that this procedure provokes organic matter enrichment of water, increasing the solid yield (SY) of the process for the same reaction conditions. In principle, this is an advantage as the quantity of hydrochar obtained for the same starting biomass charge would be increased, saving water supply. However, other aspects should be considered. For instance, if the aim of the hypothetical plant was the production of solid biochar as the main product for further commercialization, water recirculation would be interesting, but it should be taken into account the kind of biomass. It is possible that, for some kinds of biomass, process water is not suitable for its reuse in the same process, as in the case of sewage sludge, due to its high content in heavy metals. Therefore, depending on the type of biomass, recirculation could be suitable or not. Another aspect to be considered is the aim of the HTC plant. If it is going to be devoted to the production of interesting chemicals in water, the amount of these products might be varied during recirculation of water, even decreasing its concentration at the end of the process. As a consequence, chemical analysis of the liquid is another factor to be considered to assess water recirculation. On the other hand, an important detail for plant design and implementation is the calculation of recirculation systems, especially concerning pipelines, the kind of opening and closing of valves, and water inlet and outlet, with the possibility of transferring water to another reactor or system. As it can be seen, these are only some of the issues that should be taken into account regarding process water, requiring a thorough reflection about these characteristics, along with other design aspects.

5.2.2 Biomass Supply to the Reactor and Removal of Final Products

This aspect, even though apparently minor, could be a problematic issue. Firstly, it should be noted that biomass supply to the HTC reactor requires some technical considerations that allow the opening and the hermetic closing of the system. Secondly, once the reaction takes place, it requires the removal of the solid and liquid products from the reactor. Therefore, different pipeline and extraction systems are necessary to carry out this task, especially to conduct process water, as well as other systems for solid extraction from the reactor. For solid hydrochar extraction, it should be noted that depending on the kind of biomass, part of final hydrochar can be adhered to the reactor. Additionally, the possible presence of solid residue in the reactor after hydrochar removal can be problematic. As a matter of fact, the presence of biomass waste in the reactor could be unavoidable once HTC reaction takes place and after product removal. Thus, it would be interesting and advisable the implementation of a cleaning procedure to remove impurities. It is necessary to point out that the HTC plant design depends on the kind of biomass to be processed. The approach to treat sewage sludge or lignocellulosic biomass is not the same in HTC processes.

Final products will have different characteristics, apart from the fact that they depend, in turn, on the reaction conditions. As a consequence, for the correct design of a HTC plant, it is necessary to previously establish the kind of biomass that will be supplied, possibly requiring the use of multiple reactors (with different sizes or configurations) or different operating conditions. For instance, if HTC process requires the use of catalysts, even for the same kind of biomass, it would be suitable to resort to different reactors for both cases, as the operating conditions and final products will be different.

Regarding liquid extraction, an important element to be considered is the establishment of a filtering system and its post-treatment. In fact, a first step would be the incorporation of a filtering system in process water. This filtering is necessary as it would remove hydrochar particles in treatment water. After filtration, a post-treatment system would be necessary, along with handling and conservation of process water. In any case, not every HTC plant is designed to exploit the liquid phase in this process, so post-treatment of this process water could be reduced or deleted, as process water removal is the only interesting step in this case. Considering the above, it is clear that the design of biomass supply and product extraction requires a series of considerations (not only technical, but also environmental when it comes to the removal of impurity or non-necessary wastes).

5.2.3 Chemical Reactions in HTC Process

This subject has been thoroughly covered in Chap. 4. Nevertheless, it is worth noting that this is a very interesting research line in HTC. As a matter of fact, the precise knowledge of the chemical reactions that take place during this process is important in order to control the whole system during production at industrial scale. It is true that the reactions taking place in HTC are very complex, but the development of new kinetic models could considerably contribute to the optimization of HTC facilities. There are many advantages related to this fact, for instance, the better calculation of expected concentration of a specific chemical product. Another advantage could be the optimization of reactor design. Thus, for the same operating conditions, the size of a certain reactor could be reduced or properly adapted to a specific process. However, the exact knowledge of every chemical reaction taking place during HTC is unfeasible nowadays. Even so, new research about this subject along with advanced simulation systems or models could change this situation in the medium term, currently implying a promising research line.

5.2.4 Environmental Issues

Throughout this book, one of the main advantages of HTC implementation has been clearly pointed out, that is, the low gas emission related to this process, especially concerning greenhouse gas emissions. However, as previously explained, the removal of some impurities or hydrochar waste from the reactor is necessary, along with the removal of process water, which usually implies environmental challenges if these wastes are not properly managed. This way, both products can be considered as wastes in HTC process and, consequently, they should be removed in a clean way. Nowadays, the removal of treatment water can be a challenge, as this liquid does not present any utility unless the aim of a certain HTC plant is to obtain a liquid with specific characteristics. Therefore, in a plant of these characteristics, it would be necessary to establish removal protocols, including the treatment of this waste, considering and establishing standards according to current national and international laws. Similarly, recirculation water, as explained in this chapter, should be considered with a similar approach, requiring its removal at the end of recirculation stages or, at least, a treatment to make it suitable for recirculation in further processes. However, the management of these wastes should be taken into account in both cases (for process water with or without recirculation).

5.2.5 Plant Location

This is a determining factor when it comes to assess the economic feasibility of an industrial plant. Thus, the location of a HTC plant in a specific geographic area could be decisive for its economic performance, as there are important key factors to be considered concerning this subject. First of all, the kind of biomass required for the process should be considered to locate the plant, as it should be located as close as possible to these agricultural or pruning wastes. For example, if an industrial HTC process based on pruning wastes of olive trees is going to be implemented, it would be interesting the location of one or more facilities near regions where this crop is majority, implying a considerable economic relevance. A crucial aspect is the weight of supply cost of raw material in final prices of the obtained products, not only regarding economic issues, but also gas emissions. In fact, due to the increasing concern about the environment and the promotion of new clean technologies, it would be undesirable to implement HTC plants in areas far from the points where raw materials are obtained. Also, in order to make a facility of these characteristics attractive, the carbon balance during the production and use of products from HTC should be negative or neutral, that is, carbon compounds released to the atmosphere should be lower or similar to carbon fixed by biomass through photosynthesis.

Another interesting factor, related with the above, would be the fact that agriculture practices should provide sustainable raw materials. This is a generic assessment, referred to the energy obtained from biomass without specifying the kind of technology or energy use. However, it is worth explaining the sustainable nature of a HTC plant. Thus, biomass projects should be sustainable compared to water consumption, use of fertilizers, soil mineral balance, etc., requiring the assessment and assurance of biomass supply in the long run, including studies about the degree of dispersion of these resources.

Furthermore, concerning sustainability and plant location, quality of biomass plays an important role. Depending on the kind of waste or resource, both quality and location can significantly vary. Thus, if sewage sludge from a wastewater treatment plant is used, its heavy metal content should be considered, as it not only could influence the sustainability of the process from energy or environmental points of view, but also, as previously explained, would affect the quality of the final product obtained and the technology and operating conditions of the corresponding HTC plant. On the other hand, another important aspect is the cost related to the resource. Consequently, if the raw material is lignocellulosic biomass, several variables should be taken into account to assess its sustainability and quality, as intrinsic cost, cost of collection, transportation and delivery or costs related to pre-treatments.

Considering the above, it is clear that location and implementation of HTC plants in specific areas or regions have a determining influence on several factors such as sustainability and economic feasibility, which in turn impact each other.

5.3 Political and Economic Considerations

As observed in the previous section, several aspects about the implementation or performance of HTC plants at industrial scale have been covered. Nevertheless, there are other factors that present a strong influence in the transfer to industrial scale of this kind of facilities, like political and economic considerations.

5.3.1 Political Considerations

As HTC is an emerging technology with a great potential to transform specific wastes with difficult management into valuable resources such as hydrochar or fertilizers, these are perfect characteristics for the implementation of this technology in a circular economy and environmental sustainability context. Several companies, as previously explained in this chapter, have patented HTC processes that allow the recovery of carbon from organic waste and the production of solid biofuels with high calorific values, which proves the feasibility of this technique at industrial scale. In spite of regulatory challenges and the competition with established and mature renewable technologies, HTC offers innovative solutions for the suitable treatment of wet waste and the recovery of essential nutrients, which are in accordance with international goals (including the so-called sustainable development goals) about the reduction of greenhouse gas emissions. These challenges are diverse, like the following: First, there are some concerns, as well as a lack of legislation about the implementation of industrial plants using this technology. Nevertheless, this is not a major issue, as HTC could be easily adapted to existing processes, by coupling their reactions to other facilities. It should be worth noting that the lack of regulation is an important factor. There is no specific policy framework about the specific use of HTC, even though there are some countries where there are standards about the use of hydrochar produced by this process (which is not generalized). Also, it is important to point out that, although this technology has been widely studied for decades, there are no operative industrial facilities of HTC, as the majority of these facilities are pilot plants. In fact, one of the major problems for the implementation of this technology is the lack of data and real industrial development to consider HTC as a promising technology in an industrial context. Nevertheless, the implementation of HTC could be accelerated in future through education, favourable legislation and promotion of its integration in current energy scenario.

5.3.2 Economic Considerations

Energy efficiency of HTC, along with its operational flexibility and possibility of decentralization, makes this technology especially attractive for adaptative and local applications. However, the use of this technology presents several challenges as the requirement of an industrial scale proof about its feasibility and its positive comparison with mature technologies such as pyrolysis. Thus, transparency in process communication and economic results is crucial to build trust and overcome the existing reluctance in global market. Moreover, standardization of HTC products, as well as training of qualified personnel are essential to assure quality and facilitate the integration of this technology in circular economy. All these factors have a strong influence not only when it comes to getting funding, but also regarding economic return in the case of product commercialization.

To sum up, the real possibility of the implementation of HTC at industrial scale could be feasible, according to the proliferation of pilot or demonstration plants around the world, whose performance is acceptable according to production and efficiency. Nevertheless, there is still room for improvement, which will be achieved, especially at industrial scale, thanks to scientific research and engineering approach.

5.4 Questions

1. Technology readiness levels (TRLs) are:
    (a) Literally obtained from the NASA criteria
    (b) Adapted from the NASA criteria

(c) Levels of degree of applicability of a certain technology
(d) B and C are the correct

2. How many technology readiness levels are there, according to the NASA criteria?

(a) 16
(b) 12
(c) 9
(d) None of the above

3. According to the abovementioned TRLs, HTC would be classified as:

(a) TRL 7
(b) TRL 8
(c) TRL 9
(d) All of the above

4. Which continent presents the largest concentration of HTC full plants?

(a) Europe
(b) Asia
(c) North America
(d) A and B are correct

5. Regarding pilot (or demonstration) HTC plants, where are they mainly located?

(a) Europe
(b) Asia
(c) North America
(d) None of the above

Competing Interests The authors have no conflicts of interest to declare that are relevant to the content of this chapter.

References

1. NASA, Technology Readiness Levels (2024, April 30). https://www.Nasa.Gov/Directorates/Somd/Space-Communications-Navigation-Program/Technology-Readiness-Levels/
2. G. Farru, F.B. Scheufele, D. Moloeznik Paniagua, F. Keller, C. Jeong, D. Basso, Business and market analysis of hydrothermal carbonization process: roadmap toward implementation. Agronomy **14**(3) (2024). https://doi.org/10.3390/agronomy14030541
3. J.F. González, C.M. Álvez-Medina, S. Nogales-Delgado, Biogas steam reforming in wastewater treatment plants: opportunities and challenges. in *Energies*, vol. 16, issue 17 (Multidisciplinary Digital Publishing Institute (MDPI), 2023). https://doi.org/10.3390/en16176343

Solutions

This section shows the answers to the questions proposed in each chapter.

Chapter 1

1. (b) Friedrich Bergius
2. (b) Water and biomass
3. (b) Its temperature range is usually below 373.95 °C
4. (b) Hydrochar and HTC liquid
5. (d) All of the above
6. (b) The United States is the main country devoted to the scientific research in this subject

Chapter 2

1. (b) Function as a solvent and reagent
2. (b) 180–260 °C
3. (c) Hydrochars
4. (b) Dehydration
5. (b) Carbon dioxide
6. b) Dehydration
7. (b) Act as secondary reagents in later stages
8. (b) Aromatic compounds
9. (b) A solid generated due to the liquid composition of the medium
10. (b) Carbon dioxide and carbon monoxide
11. (c) Polymerisation
12. (b) More stable three-dimensional structures
13. (c) Improve the efficiency and selectivity of the process
14. (c) Colour of the catalyst
15. (b) Influences the stability and selectivity of the catalyst
16. (c) Sulphuric acid
17. (c) Acetic acid

B. Ledesma Cano et al., *Introduction to Hydrocarbonization*,
SpringerBriefs in Applied Sciences and Technology,
https://doi.org/10.1007/978-3-031-70039-2

18. (b) Facilitate dehydration and repolymerisation
19. (d) They favour decarbonylation and dehydration reactions
20. (a) Citric acid
21. (c) They improve the yield of hydrochar and reduce the emission of toxic compounds
22. (c) Improves dewatering efficiency and process competence
23. (b) It improves the structure and yield of the hydrochar
24. (b) They improve the thermal removal of biomass and reduce the nitrogen content in the hydrochar
25. (c) They improve the final quality of the hydrochar
26. (a) Zinc chloride ($ZnCl_2$)

Chapter 3

1. (a) In a certain process, gas released to the atmosphere is always considered a waste
2. (a) Its moisture level is considerable, which could reduce water addition in this process
3. (d) b and c are correct
4. (d) a and b are correct
5. (a) Biomass-biomass
6. (d) a and c are correct
7. (c) There is a synergistic effect

Chapter 4

Exercises

1. The following data are available:

Biomasa seca inicial, g	50
Agua añadida al proceso, g	100
HC, g	28
PW, g	15
PG	0

Therefore, the solid yield obtained is calculated as follows:

$$SY\ (\%) = \left(\frac{\text{mass of HC}}{\text{total initial mass of biomass}} \right) \times 100 = \left(\frac{28}{50} \right) \times 100 = 56\%$$

The liquid yield is calculated as follows:

$$SL\ (\%) = \left(\frac{\text{mass of PW}}{\text{total initial mass of water}} \right) \times 100 = \left(\frac{15}{100} \right) \times 100 = 15\%$$

2. From the FT-IR spectrum of an unknown sample it can be roughly observed that the position and shape of the band at 3438 cm^{-1} is due to O–H stretching vibration and may be related to the presence of water in the sample. The peak at 2920 cm^{-1} indicates asymmetric C–H elongation in methyl and methylene groups, while the absorption at 2853 cm^{-1} could be associated with symmetric stretching-type vibrations of these groups. The bands at 1730 cm^{-1} and 1633 cm^{-1} are due to the C=O stretching vibration of the conjugated and unconjugated carbonyl groups with aromatic rings, respectively.

3.

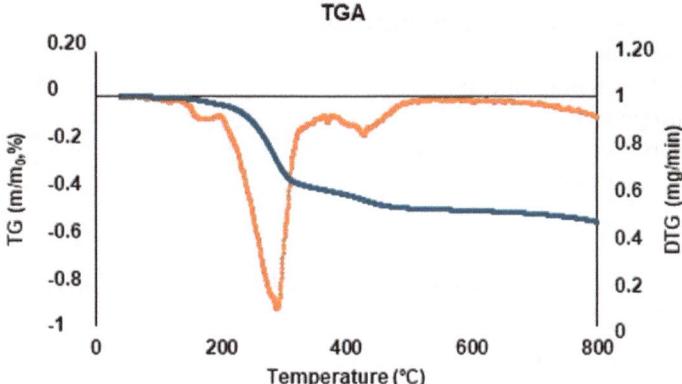

(a) The main steps in the thermal decomposition of the sample are as follows:

Temperature range, °C	Thermal decomposition stages
25–150	Dehydration
150–350	Hemicellulose decomposition
350–450	Cellulose decomposition
450–800	Lignin decomposition

(b) For the determination of the temperature at which the maximum weight loss is observed, the derivative of the TGA curve should be analysed to find the maximum peak.

Questions

The temperature at which the maximum mass loss is observed from the TGA graph is approximately 290 °C.

1. (b) To reduce the particle size and thereby improve the homogeneity and efficiency of the process

2. (c) Because it is necessary to maintain the solid and liquid water phases under controlled conditions of temperature and pressure
3. (d) Spectrophotometer
4. (b) The amount of heat released during complete combustion per unit mass.
5. (b) By difference with the total C, H, N, and S content
6. (c) Hydrochar
7. (a) FC = [100 − (Moisture + Ash + Volatile Matter)]
8. (b) Fibre analysis
9. (b) pH meter
10. (b) SEM (Scanning Electron Microscopy)
11. (b) For organic and inorganic materials
12. (b) Chemical composition
13. (c) O–H stretching (alcohol)
14. (c) Nitrogen
15. (c) Micropores
16. (c) Microporous

Chapter 5

1. (d) B and C are correct (Adapted from the NASA criteria, Levels of degree of applicability of a certain technology)
2. (c) 9
3. (d) All of the above
4. (a) Europe
5. (a) Europe